SpringerBriefs in Computer Science

T0255910

More information about this series at http://www.springer.com/series/10028

Rodrigo C. Barros · André C.P.L.F. de Carvalho
Alex A. Freitas

Automatic Design
of Decision-Tree
Induction Algorithms

Springer

Rodrigo C. Barros
Faculdade de Informática
Pontifícia Universidade Católica do Rio
  Grande do Sul
Porto Alegre, RS
Brazil

André C.P.L.F. de Carvalho
Instituto de Ciências Matemáticas e de
  Computação
Universidade de São Paulo
São Carlos, SP
Brazil

Alex A. Freitas
School of Computing
University of Kent
Canterbury, Kent
UK

ISSN 2191-5768   ISSN 2191-5776   (electronic)
SpringerBriefs in Computer Science
ISBN 978-3-319-14230-2   ISBN 978-3-319-14231-9   (eBook)
DOI 10.1007/978-3-319-14231-9

Library of Congress Control Number: 2014960035

Springer Cham Heidelberg New York Dordrecht London

Printed on acid-free paper

Springer International Publishing AG Switzerland is part of Springer Science+Business Media
(www.springer.com)

This book is dedicated to my family:
Alessandra, my wife;
Marta and Luís Fernando, my parents;
Roberta, my sister; Gael, my godson;
Lygia, my grandmother.

Rodrigo C. Barros

To Valeria, my wife, and to Beatriz,
Gabriela and Mariana, my daughters.

André C.P.L.F. de Carvalho

To Jie, my wife.

Alex A. Freitas

Contents

Notations

T	A decision tree
$\mathbf{X}$	A set of instances
N_x	The number of instances in $\mathbf{X}$, i.e., $|\mathbf{X}|$
$\mathbf{x}^j$	An instance—n-dimensional attribute vector $[x_1^j, x_2^j, \ldots, x_n^j]$—from $\mathbf{X}$, $j = 1, 2, \ldots, N_x$
$\mathbf{X_t}$	A set of instances that reach node t
A	The set of n predictive (independent) attributes $\{a_1, a_2, \ldots, a_n\}$
y	The target (class) attribute
Y	The set of k class labels $\{y_1, \ldots, y_k\}$ (or k distinct values if y is continuous)
$y(x)$	Returns the class label (or target value) of instance $\mathbf{x} \in \mathbf{X}$
$a_i(x)$	Returns the value of attribute a_i from instance $\mathbf{x} \in \mathbf{X}$
$dom(a_i)$	The set of values attribute a_i can take
$|a_i|$	The number of partitions resulting from splitting attribute a_i
$\mathbf{X_{a_i=v_j}}$	The set of instances in which attribute a_i takes a value contemplated by partition v_j. Edge v_j can refer to a nominal value, to a set of nominal values, or even to a numeric interval
$N_{v_j, \bullet}$	The number of instances in which attribute a_i takes a value contemplated by partition v_j, i.e., $|\mathbf{X_{a_i=v_j}}|$
$\mathbf{X_{y=y_l}}$	The set of instances in which the class attribute takes the label (value) y_l
$N_{\bullet, y_l}$	The number of instances in which the class attribute takes the label (value) y_l, i.e., $|\mathbf{X_{y=y_l}}|$
$N_{v_j \cap y_l}$	The number of instances in which attribute a_i takes a value contemplated by partition v_j and in which the target attribute takes the label (value) y_l
v_X	The target (class) vector $[N_{\bullet, y_1}, \ldots, N_{\bullet, y_k}]$ associated to $\mathbf{X}$
p_y	The target (class) probability vector $[p_{\bullet, y_1}, \ldots, p_{\bullet, y_k}]$
$p_{\bullet, y_l}$	The estimated probability of a given instance belonging to class y_l, i.e., $\frac{N_{\bullet, y_l}}{N_x}$

$p_{v_j,\bullet}$ The estimated probability of a given instance being contemplated by partition v_j, i.e., $\frac{N_{v_j,\bullet}}{N_x}$

$p_{v_j \cap y_l}$ The estimated joint probability of a given instance being contemplated by partition v_j and also belonging to class y_l, i.e., $\frac{N_{v_j \cap y_l}}{N_x}$

$p_{y_l|v_j}$ The conditional probability of a given instance belonging to class y_l given that it is contemplated by partition v_j, i.e., $\frac{N_{v_j \cap y_l}}{N_{v_j,\bullet}}$

$p_{v_j|y_l}$ The conditional probability of a given instance being contemplated by partition v_j given that it belongs to class y_l, i.e., $\frac{N_{v_j \cap y_l}}{N_{\bullet,y_l}}$

ζ_T The set of nonterminal nodes in decision tree T

λ_T The set of terminal nodes in decision tree T

$\aleph_T$ The set of nodes in decision tree T, i.e., $\aleph_T = \zeta_T \cup \lambda_T$

$T^{(t)}$ A (sub)tree rooted in node t

$E^{(t)}$ The number of instances in t that do not belong to the majority class of that node

Chapter 1
Introduction

Classification, which is the data mining task of assigning objects to predefined categories, is widely used in the process of intelligent decision making. Many classification techniques have been proposed by researchers in machine learning, statistics, and pattern recognition. Such techniques can be roughly divided according to the their level of comprehensibility. For instance, techniques that produce interpretable classification models are known as *white-box* approaches, whereas those that do not are known as *black-box* approaches. There are several advantages in employing white-box techniques for classification, such as increasing the user confidence in the prediction, providing new insight about the classification problem, and allowing the detection of errors either in the model or in the data [12]. Examples of white-box classification techniques are classification rules and decision trees. The latter is the main focus of this book.

A decision tree is a classifier represented by a flowchart-like tree structure that has been widely used to represent classification models, specially due to its comprehensible nature that resembles the human reasoning. In a recent poll from the *kdnuggets* website [13], decision trees figured as the most used data mining/analytic method by researchers and practitioners, reaffirming its importance in machine learning tasks. Decision-tree induction algorithms present several advantages over other learning algorithms, such as robustness to noise, low computational cost for generating the model, and ability to deal with redundant attributes [22].

Several attempts on optimising decision-tree algorithms have been made by researchers within the last decades, even though the most successful algorithms date back to the mid-80s [4] and early 90s [21]. Many strategies were employed for deriving accurate decision trees, such as bottom-up induction [1, 17], linear programming [3], hybrid induction [15], and ensemble of trees [5], just to name a few. Nevertheless, no strategy has been more successful in generating accurate and comprehensible decision trees with low computational effort than the greedy top-down induction strategy.

A greedy top-down decision-tree induction algorithm recursively analyses if a sample of data should be partitioned into subsets according to a given rule, or if no further partitioning is needed. This analysis takes into account a stopping criterion, for

© The Author(s) 2015   1
R.C. Barros et al., *Automatic Design of Decision-Tree Induction Algorithms*,
SpringerBriefs in Computer Science, DOI 10.1007/978-3-319-14231-9_1

deciding when tree growth should halt, and a splitting criterion, which is responsible for choosing the "best" rule for partitioning a subset. Further improvements over this basic strategy include pruning tree nodes for enhancing the tree's capability of dealing with noisy data, and strategies for dealing with missing values, imbalanced classes, oblique splits, among others.

A very large number of approaches were proposed in the literature for each one of these *design components* of decision-tree induction algorithms. For instance, new measures for node-splitting tailored to a vast number of application domains were proposed, as well as many different strategies for selecting multiple attributes for composing the node rule (multivariate split). There are even studies in the literature that survey the numerous approaches for pruning a decision tree [6, 9]. It is clear that by improving these design components, more effective decision-tree induction algorithms can be obtained.

An approach that has been increasingly used in academia is the induction of decision trees through evolutionary algorithms (EAs). They are essentially algorithms inspired by the principle of natural selection and genetics. In nature, individuals are continuously evolving, adapting to their living environment. In EAs, each "individual" represents a candidate solution to the target problem. Each individual is evaluated by a fitness function, which measures the quality of its corresponding candidate solution. At each generation, the best individuals have a higher probability of being selected for reproduction. The selected individuals undergo operations inspired by genetics, such as crossover and mutation, producing new offspring which will replace the parents, creating a new generation of individuals. This process is iteratively repeated until a stopping criterion is satisfied [8, 11]. Instead of local search, EAs perform a robust global search in the space of candidate solutions. As a result, EAs tend to cope better with attribute interactions than greedy methods [10].

The number of EAs for decision-tree induction has grown in the past few years, mainly because they report good predictive performance whilst keeping the comprehensibility of decision trees [2]. In this approach, each individual of the EA is a decision tree, and the evolutionary process is responsible for searching the solution space for the "near-optimal" tree regarding a given data set. A disadvantage of this approach is that it generates a decision tree tailored to a single data set. In other words, an EA has to be executed every time we want to induce a tree for a giving data set. Since the computational effort of executing an EA is much higher than executing the traditional greedy approach, it may not be the best strategy for inducing decision trees in time-constrained scenarios.

Whether we choose to induce decision trees through the greedy strategy (top-down, bottom-up, hybrid induction), linear programming, EAs, ensembles, or any other available method, we are susceptible to the method's inductive bias. Since we know that certain inductive biases are more suitable to certain problems, and that no method is best for every single problem (i.e., the no free lunch theorem [26]), there is a growing interest in developing automatic methods for deciding which learner to use in each situation. A whole new research area named *meta-learning* has emerged for solving this problem [23]. Meta-learning is an attempt to understand data *a priori* of executing a learning algorithm. In a particular branch of meta-learning, *algorithm recommendation*, data that describe the characteristics of data sets and learning algorithms (i.e., meta-data) are collected, and a learning algorithm is employed to interpret these meta-data and suggest a particular learner (or ranking a few learners) in order to better solve the problem at hand. Meta-learning has a few limitations, though. For instance, it provides a limited number of algorithms to be selected from a list. In addition, it is not an easy task to define the set of meta-data that will hopefully contain useful information for identifying the best algorithm to be employed.

For avoiding the limitations of traditional meta-learning approaches, a promising idea is to automatically develop algorithms tailored to a given domain or to a specific set of data sets. This approach can be seen as a particular type of meta-learning, since we are learning the "optimal learner" for specific scenarios. One possible technique for implementing this idea is *genetic programming* (GP). It is a branch of EAs that arose as a paradigm for evolving computer programs in the beginning of the 90s [16]. The idea is that each individual in GP is a computer program that evolves during the evolutionary process of the EA. Hopefully, at the end of evolution, GP will have found the appropriate algorithm (best individual) for the problem we want to solve. Pappa and Freitas [20] cite two examples of EA applications in which the evolved individual outperformed the best human-designed solution for the problem. In the first application [14], the authors designed a simple satellite dish holder boom (connection between the satellite's body and the communication dish) using an EA. This automatically designed dish holder boom, albeit its bizarre appearance, was shown to be 20,000 % better than the human-designed shape. The second application [18] was concerning the automatic discovery of a new form of boron (chemical element). There are only four known forms of borons, and the last one was discovered by an EA.

A recent research area within the combinatorial optimisation field named "hyper-heuristics" (HHs) has emerged with a similar goal: searching in the heuristics space, or in other words, *heuristics to choose heuristics* [7]. HHs are related to metaheuristics, though with the difference that they operate on a search space of heuristics whereas metaheuristics operate on a search space of solutions to a given problem. Nevertheless, HHs usually employ metaheuristics (e.g., evolutionary algorithms) as the search methodology to look for suitable heuristics to a given problem [19]. Considering that an algorithm or its components can be seen as heuristics, one may say that HHs are also suitable tools to automatically design custom (tailor-made) algorithms.

Whether we name it "an EA for automatically designing algorithms" or "hyper-heuristics", in both cases there is a set of human designed components or heuristics, surveyed from the literature, which are chosen to be the starting point for the evolutionary process. The expected result is the automatic generation of new procedural components and heuristics during evolution, depending of course on which components are provided to the EA and the respective "freedom" it has for evolving the solutions.

The automatic design of complex algorithms is a much desired task by researchers. It was envisioned in the early days of artificial intelligence research, and more recently has been addressed by machine learning and evolutionary computation research groups [20, 24, 25]. Automatically designing machine learning algorithms can be seen as the task of teaching the computer how to create programs that learn from experience. By providing an EA with initial human-designed programs, the evolutionary process will be in charge of generating new (and possibly better) algorithms for the problem at hand. Having said that, we believe an EA for automatically discovering new decision-tree induction algorithms may be the solution to avoid the drawbacks of the current decision-tree approaches, and this is going to be the main topic of this book.

1.1 Book Outline

This book is structured in 7 chapters, as follows.

Chapter 2 **[Decision-Tree Induction]**. This chapter presents the origins, basic concepts, detailed components of top-down induction, and also other decision-tree induction strategies.

Chapter 3 **[Evolutionary Algorithms and Hyper-Heuristics]**. This chapter covers the origins, basic concepts, and techniques for both Evolutionary Algorithms and Hyper-Heuristics.

Chapter 4 **[HEAD-DT: Automatic Design of Decision-Tree Induction Algorithms]**. This chapter introduces and discusses the hyper-heuristic evolutionary algorithm that is capable of automatically designing decision-tree algorithms. Details such as the evolutionary scheme, building blocks, fitness evaluation, selection, genetic operators, and search space are covered in depth.

Chapter 5 **[HEAD-DT: Experimental Analysis]**. This chapter presents a thorough empirical analysis on the distinct scenarios in which HEAD-DT may be applied to. In addition, a discussion on the cost effectiveness of automatic design, as well as examples of automatically-designed algorithms and a baseline comparison between genetic and random search are also presented.

Chapter 6 **[HEAD-DT: Fitness Function Analysis]**. This chapter conducts an investigation of 15 distinct versions for HEAD-DT by varying its fitness function, and a new set of experiments with the best-performing strategies in balanced and imbalanced data sets is described.

Chapter 7 **[Conclusions]**. We finish this book by presenting the current limitations of the automatic design, as well as our view of several exciting opportunities for future work.

References

1. R.C. Barros et al., A bottom-up oblique decision tree induction algorithm, in *11th International Conference on Intelligent Systems Design and Applications*. pp. 450–456 (2011)
2. R.C. Barros et al., A survey of evolutionary algorithms for decision-tree induction. IEEE Trans. Syst., Man, Cybern., Part C: Appl. Rev. **42**(3), 291–312 (2012)
3. K. Bennett, O. Mangasarian, Multicategory discrimination via linear programming. Optim. Methods Softw. **2**, 29–39 (1994)
4. L. Breiman et al., *Classification and Regression Trees* (Wadsworth, Belmont, 1984)
5. L. Breiman, Random forests. Mach. Learn. **45**(1), 5–32 (2001)
6. L. Breslow, D. Aha, Simplifying decision trees: a survey. Knowl. Eng. Rev. **12**(01), 1–40 (1997)
7. P. Cowling, G. Kendall, E. Soubeiga, A Hyperheuristic Approach to Scheduling a Sales Summit, in *Practice and Theory of Automated Timetabling III*, Vol. 2079. Lecture Notes in Computer Science, ed. by E. Burke, W. Erben (Springer, Berlin, 2001), pp. 176–190
8. A.E. Eiben, J.E. Smith, *Introduction to Evolutionary Computing (Natural Computing Series)* (Springer, Berlin, 2008)
9. F. Esposito, D. Malerba, G. Semeraro, A comparative analysis of methods for pruning decision trees. IEEE Trans. Pattern Anal. Mach. Intell. **19**(5), 476–491 (1997)
10. A.A. Freitas, *Data Mining and Knowledge Discovery with Evolutionary Algorithms* (Springer, New York, 2002). ISBN: 3540433317
11. A.A. Freitas, A Review of evolutionary Algorithms for Data Mining, in *Soft Computing for Knowledge Discovery and Data Mining*, ed. by O. Maimon, L. Rokach (Springer, Berlin, 2008), pp. 79–111. ISBN: 978-0-387-69935-6
12. A.A. Freitas, D.C. Wieser, R. Apweiler, On the importance of comprehensible classification models for protein function prediction. IEEE/ACM Trans. Comput. Biol. Bioinform. **7**, 172–182 (2010). ISSN: 1545–5963
13. KDNuggets, *Poll: Data mining/analytic methods you used frequently in the past 12 months* (2007)
14. A. Keane, S. Brown, The design of a satellite boom with enhanced vibration performance using genetic algorithm techniques, in *Conference on Adaptative Computing in Engineering Design and Control*. Plymouth, pp. 107–113 (1996)
15. B. Kim, D. Landgrebe, Hierarchical classifier design in high-dimensional numerous class cases. IEEE Trans. Geosci. Remote Sens. **29**(4), 518–528 (1991)
16. J.R. Koza, *Genetic Programming: On the Programming of Computers by Means of Natural Selection* (MIT Press, Cambridge, 1992). ISBN: 0-262-11170-5
17. G. Landeweerd et al., Binary tree versus single level tree classification of white blood cells. Pattern Recognit. **16**(6), 571–577 (1983)
18. A.R. Oganov et al., Ionic high-pressure form of elemental boron. Nature **457**, 863–867 (2009)
19. G.L. Pappa et al., Contrasting meta-learning and hyper-heuristic research: the role of evolutionary algorithms, in *Genetic Programming and Evolvable Machines* (2013)
20. G.L. Pappa, A.A. Freitas, *Automating the Design of Data Mining Algorithms: An Evolutionary Computation Approach* (Springer Publishing Company Incorporated, New York, 2009)
21. J.R. Quinlan, *C4.5: Programs for Machine Learning* (Morgan Kaufmann, San Francisco, 1993). ISBN: 1-55860-238-0
22. L. Rokach, O. Maimon, Top-down induction of decision trees classifiers—a survey. IEEE Trans. Syst. Man, Cybern. Part C: Appl. Rev. **35**(4), 476–487 (2005)
23. K.A. Smith-Miles, Cross-disciplinary perspectives on meta-learning for algorithm selection. ACM Comput. Surv. **41**, 6:1–6:25 (2009)
24. K.O. Stanley, R. Miikkulainen, Evolving neural networks through augmenting topologies. Evol. Comput. **10**(2), 99–127 (2002). ISSN: 1063–6560
25. A. Vella, D. Corne, C. Murphy, Hyper-heuristic decision tree induction, in *World Congress on Nature and Biologically Inspired Computing*, pp. 409–414 (2010)
26. D.H. Wolpert, W.G. Macready, No free lunch theorems for optimization. IEEE Trans. Evol. Comput. **1**(1), 67–82 (1997)

Chapter 2
Decision-Tree Induction

Abstract Decision-tree induction algorithms are highly used in a variety of domains for knowledge discovery and pattern recognition. They have the advantage of producing a comprehensible classification/regression model and satisfactory accuracy levels in several application domains, such as medical diagnosis and credit risk assessment. In this chapter, we present in detail the most common approach for decision-tree induction: top-down induction (Sect. 2.3). Furthermore, we briefly comment on some alternative strategies for induction of decision trees (Sect. 2.4). Our goal is to summarize the main design options one has to face when building decision-tree induction algorithms. These design choices will be specially interesting when designing an evolutionary algorithm for evolving decision-tree induction algorithms.

Keywords Decision trees · Hunt's algorithm · Top-down induction · Design components

2.1 Origins

Automatically generating rules in the form of decision trees has been object of study of most research fields in which data exploration techniques have been developed [78]. Disciplines like engineering (pattern recognition), statistics, decision theory, and more recently artificial intelligence (machine learning) have a large number of studies dedicated to the generation and application of decision trees.

In statistics, we can trace the origins of decision trees to research that proposed building binary segmentation trees for understanding the relationship between target and input attributes. Some examples are AID [107], MAID [40], THAID [76], and CHAID [55]. The application that motivated these studies is survey data analysis. In engineering (pattern recognition), research on decision trees was motivated by the need to interpret images from remote sensing satellites in the 70s [46]. Decision trees, and induction methods in general, arose in machine learning to avoid the knowledge acquisition bottleneck for expert systems [78].

Specifically regarding top-down induction of decision trees (by far the most popular approach of decision-tree induction), Hunt's Concept Learning System (CLS)

© The Author(s) 2015

R.C. Barros et al., *Automatic Design of Decision-Tree Induction Algorithms*,
SpringerBriefs in Computer Science, DOI 10.1007/978-3-319-14231-9_2

[49] can be regarded as the pioneering work for inducing decision trees. Systems that directly descend from Hunt's CLS are ID3 [91], ACLS [87], and Assistant [57].

2.2 Basic Concepts

Decision trees are an efficient nonparametric method that can be applied either to classification or to regression tasks. They are hierarchical data structures for supervised learning whereby the input space is split into local regions in order to predict the dependent variable [2].

A decision tree can be seen as a graph $G = (V, E)$ consisting of a finite, nonempty set of nodes (vertices) V and a set of edges E. Such a graph has to satisfy the following properties [101]:

- The edges must be ordered pairs (v, w) of vertices, i.e., the graph must be directed;
- There can be no cycles within the graph, i.e., the graph must be acyclic;
- There is exactly one node, called the root, which no edges enter;
- Every node, except for the root, has exactly one entering edge;
- There is a unique path—a sequence of edges of the form $(v_1, v_2), (v_2, v_3), \ldots,$ (v_{n-1}, v_n)—from the root to each node;
- When there is a path from node v to w, $v \neq w$, v is a proper *ancestor* of w and w is a proper *descendant* of v. A node with no proper descendant is called a *leaf* (or a *terminal*). All others are called *internal* nodes (except for the root).

Root and internal nodes hold a test over a given data set attribute (or a set of attributes), and the edges correspond to the possible outcomes of the test. Leaf nodes can either hold class labels (classification), continuous values (regression), (non-) linear models (regression), or even models produced by other machine learning algorithms. For predicting the dependent variable value of a certain instance, one has to navigate through the decision tree. Starting from the root, one has to follow the edges according to the results of the tests over the attributes. When reaching a leaf node, the information it contains is responsible for the prediction outcome. For instance, a traditional decision tree for classification holds class labels in its leaves.

Decision trees can be regarded as a disjunction of conjunctions of constraints on the attribute values of instances [74]. Each path from the root to a leaf is actually a conjunction of attribute tests, and the tree itself allows the choice of different paths, that is, a disjunction of these conjunctions.

Other important definitions regarding decision trees are the concepts of *depth* and *breadth*. The average number of layers (levels) from the root node to the terminal nodes is referred to as the *average depth* of the tree. The average number of internal nodes in each level of the tree is referred to as the *average breadth* of the tree. Both depth and breadth are indicators of tree complexity, that is, the higher their values are, the more complex the corresponding decision tree is.

In Fig. 2.1, an example of a general decision tree for classification is presented. Circles denote the root and internal nodes whilst squares denote the leaf nodes. In

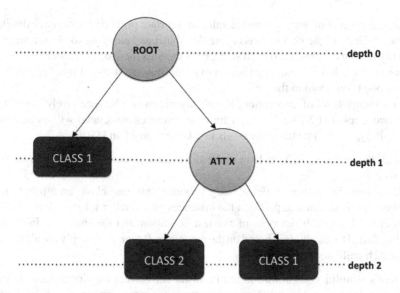

Fig. 2.1 Example of a general decision tree for classification

this particular example, the decision tree is designed for classification and thus the leaf nodes hold class labels.

There are many decision trees that can be grown from the same data. Induction of an optimal decision tree from data is considered to be a hard task. For instance, Hyafil and Rivest [50] have shown that constructing a minimal binary tree with regard to the expected number of tests required for classifying an unseen object is an NP-complete problem. Hancock et al. [43] have proved that finding a minimal decision tree consistent with the training set is NP-Hard, which is also the case of finding the minimal equivalent decision tree for a given decision tree [129], and building the optimal decision tree from decision tables [81]. These papers indicate that growing optimal decision trees (a brute-force approach) is only feasible in very small problems.

Hence, it was necessary the development of heuristics for solving the problem of growing decision trees. In that sense, several approaches which were developed in the last three decades are capable of providing reasonably accurate, if suboptimal, decision trees in a reduced amount of time. Among these approaches, there is a clear preference in the literature for algorithms that rely on a greedy, top-down, recursive partitioning strategy for the growth of the tree (top-down induction).

2.3 Top-Down Induction

Hunt's Concept Learning System framework (CLS) [49] is said to be the pioneer work in top-down induction of decision trees. CLS attempts to minimize the cost of classifying an object. Cost, in this context, is referred to two different concepts: the measurement cost of determining the value of a certain property (attribute) exhibited by the object, and the cost of classifying the object as belonging to class j when it actually belongs to class k. At each stage, CLS exploits the space of possible decision trees to a fixed depth, chooses an action to minimize cost in this limited space, then moves one level down in the tree.

In a higher level of abstraction, Hunt's algorithm can be recursively defined in only two steps. Let $\mathbf{X}_t$ be the set of training instances associated with node t and $y = \{y_1, y_2, \ldots, y_k\}$ be the class labels in a k-class problem [110]:

1. If all the instances in $\mathbf{X}_t$ belong to the same class y_t then t is a leaf node labeled as y_t
2. If $\mathbf{X}_t$ contains instances that belong to more than one class, an attribute test condition is selected to partition the instances into smaller subsets. A child node is created for each outcome of the test condition and the instances in $\mathbf{X}_t$ are distributed to the children based on the outcomes. Recursively apply the algorithm to each child node.

Hunt's simplified algorithm is the basis for all current top-down decision-tree induction algorithms. Nevertheless, its assumptions are too stringent for practical use. For instance, it would only work if every combination of attribute values is present in the training data, and if the training data is inconsistency-free (each combination has a unique class label).

Hunt's algorithm was improved in many ways. Its stopping criterion, for example, as expressed in step 1, requires all leaf nodes to be pure (i.e., belonging to the same class). In most practical cases, this constraint leads to enormous decision trees, which tend to suffer from *overfitting* (an issue discussed later in this chapter). Possible solutions to overcome this problem include prematurely stopping the tree growth when a minimum level of impurity is reached, or performing a *pruning* step after the tree has been fully grown (more details on other stopping criteria and on pruning in Sects. 2.3.2 and 2.3.3). Another design issue is how to select the attribute test condition to partition the instances into smaller subsets. In Hunt's original approach, a cost-driven function was responsible for partitioning the tree. Subsequent algorithms such as ID3 [91, 92] and C4.5 [89] make use of information theory based functions for partitioning nodes in purer subsets (more details on Sect. 2.3.1).

An up-to-date algorithmic framework for top-down induction of decision trees is presented in [98], and we reproduce it in Algorithm 1. It contains three procedures: one for growing the tree (*treeGrowing*), one for pruning the tree (*treePruning*) and one to combine those two procedures (*inducer*). The first issue to be discussed is how to select the test condition $f(A)$, i.e., how to select the best combination of attribute(s) and value(s) for splitting nodes.

Algorithm 1 Generic algorithmic framework for top-down induction of decision trees. Inputs are the training set **X**, the predictive attribute set A and the target attribute y.

1: **procedure** *inducer*(**X**, A, y)
2:       $T = $ *treeGrowing*(**X**, A, y)
3:       **return** *treePruning*(**X**, T)
4: **end procedure**
5: **procedure** *treeGrowing*(**X**, A, y)
6:       Create a tree T
7:       **if** one of the stopping criteria is fulfilled **then**
8:             Mark the root node in T as a leaf with the most common value of y in **X**
9:       **else**
10:            Find an attribute test condition $f(A)$ such that splitting **X** according to $f(A)$'s outcomes $(v_1, \ldots, v_l)$ yields
                   the best splitting measure value
11:            **if** best splitting measure value $>$ *threshold* **then**
12:                   Label the root node in T as $f(A)$
13:                   **for** each outcome v_i of $f(A)$ **do**
14:                         $\mathbf{X}_{f(A)=v_i} = \{x \in \mathbf{X} \mid f(A) = v_i\}$
15:                         $Subtree_i = $ *treeGrowing*($\mathbf{X}_{f(A=v_i)}$, A, y)
16:                         Connect the root node of T to $Subtree_i$ and label the corresponding edge as v_i
17:                   **end for**
18:            **else**
19:                   Mark the root node of T as a leaf and label it as the most common value of y in **X**
20:            **end if**
21:       **end if** **return** T
22: **end procedure**
23: **procedure** *treePruning*(**X**, T)
24:       **repeat**
25:             Select a node t in T such that pruning it maximally improves some evaluation criterion
26:             **if** $T \neq \emptyset$ **then**
27:                   $T = $ *pruned*(T, t)
28:             **end if**
29:       **until** $T = \emptyset$ **return** T
30: **end procedure**

2.3.1 Selecting Splits

A major issue in top-down induction of decision trees is which attribute(s) to choose for splitting a node in subsets. For the case of *axis-parallel* decision trees (also known as *univariate*), the problem is to choose the attribute that better discriminates the input data. A decision rule based on such an attribute is thus generated, and the input data is filtered according to the outcomes of this rule. For *oblique* decision trees (also known as *multivariate*), the goal is to find a combination of attributes with good discriminatory power. Either way, both strategies are concerned with ranking attributes quantitatively.

We have divided the work in univariate criteria in the following categories: (i) information theory-based criteria; (ii) distance-based criteria; (iii) other classification criteria; and (iv) regression criteria. These categories are sometimes fuzzy and do not constitute a taxonomy by any means. Many of the criteria presented in a given category can be shown to be approximations of criteria in other categories.

2.3.1.1 Information Theory-Based Criteria

Examples of this category are criteria based, directly or indirectly, on Shannon's entropy [104]. Entropy is known to be a unique function which satisfies the four axioms of uncertainty. It represents the average amount of information when coding each class into a codeword with ideal length according to its probability. Some interesting facts regarding entropy are:

• For a fixed number of classes, entropy increases as the probability distribution of classes becomes more uniform;
• If the probability distribution of classes is uniform, entropy increases logarithmically as the number of classes in a sample increases;
• If a partition induced on a set $\mathbf{X}$ by an attribute a_j is a refinement of a partition induced by a_i, then the entropy of the partition induced by a_j is never higher than the entropy of the partition induced by a_i (and it is only equal if the class distribution is kept identical after partitioning). This means that progressively refining a set in sub-partitions will continuously decrease the entropy value, regardless of the class distribution achieved after partitioning a set.

The first splitting criterion that arose based on entropy is the *global mutual information* (GMI) [41, 102, 108], given by:

$$GMI(a_i, \mathbf{X}, y) = \frac{1}{N_x} \sum_{l=1}^{k} \sum_{j=1}^{|a_i|} N_{v_j \cap y_l} \log_e \frac{N_{v_j \cap y_l} N_x}{N_{v_j, \bullet} N_{\bullet, y_l}} \qquad (2.1)$$

Ching et al. [22] propose the use of GMI as a tool for supervised discretization. They name it *class-attribute mutual information*, though the criterion is exactly the same. GMI is bounded by zero (when a_i and y are completely independent) and its maximum value is $max(\log_2 |a_i|, \log_2 k)$ (when there is a maximum correlation between a_i and y). Ching et al. [22] reckon this measure is biased towards attributes with many distinct values, and thus propose the following normalization called *class-attribute interdependence redundancy* (CAIR):

$$CAIR(a_i, \mathbf{X}, y) = \frac{GMI}{-\sum_{j=1}^{|a_i|} \sum_{l=1}^{k} p_{v_j \cap y_l} \log_2 p_{v_j \cap y_l}} \qquad (2.2)$$

which is actually dividing GMI by the joint entropy of a_i and y. Clearly *CAIR* $(a_i, \mathbf{X}, y) \geq 0$, since both GMI and the joint entropy are greater (or equal) than zero. In fact, $0 \leq CAIR(a_i, \mathbf{X}, y) \leq 1$, with $CAIR(a_i, \mathbf{X}, y) = 0$ when a_i and y are totally independent and $CAIR(a_i, \mathbf{X}, y) = 1$ when they are totally dependent. The term *redundancy* in CAIR comes from the fact that one may discretize a continuous attribute in intervals in such a way that the class-attribute interdependence is kept intact (i.e., *redundant* values are combined in an interval). In the decision tree partitioning context, we must look for an attribute that maximizes CAIR (or similarly, that maximizes GMI).

Information gain [18, 44, 92, 122] is another example of measure based on Shannon's entropy. It belongs to the class of the so-called *impurity-based criteria*. The term *impurity* refers to the level of class separability among the subsets derived from a split. A *pure* subset is the one whose instances belong all to the same class. Impurity-based criteria are usually measures with values in [0, 1] where 0 refers to the purest subset possible and 1 the impurest (class values are equally distributed among the subset instances). More formally, an impurity-based criterion $\phi(.)$ presents the following properties:

- $\phi(.)$ is minimum if $\exists i$ such that $p_{\bullet,y_i} = 1$;
- $\phi(.)$ is maximum if $\forall i, 1 \leq i \leq k, p_{\bullet,y_i} = 1/k$;
- $\phi(.)$ is symmetric with respect to components of p_y;
- $\phi(.)$ is smooth (differentiable everywhere) in its range.

Note that impurity-based criteria tend to favor a particular split for which, on average, the class distribution in each subset is most uneven. The impurity is measured before and after splitting a node according to each possible attribute. The attribute which presents the greater gain in purity, i.e., that maximizes the difference of impurity taken before and after splitting the node, is chosen. The gain in purity $(\Delta\Phi)$ can be defined as:

$$\Delta\Phi(a_i, \mathbf{X}, y) = \phi(y, \mathbf{X}) - \sum_{j=1}^{|a_i|} p_{v_j,\bullet} \times \phi(y, \mathbf{X}_{\mathbf{a_i}=\mathbf{v_j}}) \qquad (2.3)$$

The goal of information gain is to maximize the reduction in entropy due to splitting each individual node. Entropy can be defined as:

$$\phi^{entropy}(\mathbf{X}, y) = - \sum_{l=1}^{k} p_{\bullet,y_l} \times \log_2 p_{\bullet,y_l}. \qquad (2.4)$$

If entropy is calculated in (2.3), then $\Delta\Phi(a_i, \mathbf{X})$ is the information gain measure, which calculates the goodness of splitting the instance space $\mathbf{X}$ according to the values of attribute a_i.

Wilks [126] has proved that as $N \rightarrow \infty$, $2 \times N_x \times GMI(a_i, \mathbf{X}, y)$ (or similarly replacing GMI by information gain) approximate the χ^2 distribution. This measure is often regarded as the *G statistics* [72, 73]. White and Liu [125] point out that the G statistics should be adjusted since the work of Mingers [72] uses logarithms to base e, instead of logarithms to base 2. The adjusted G statistics is given by $2 \times N_x \times \Delta\Phi^{IG} \times \log_e 2$. Instead of using the value of this measure as calculated, we can compute the probability of such a value occurring from the χ^2 distribution on the assumption that there is no association between the attribute and the classes. The higher the calculated value, the less likely it is to have occurred given the assumption. The advantage of using such a measure is making use of the levels of significance it provides for deciding whether to include an attribute at all.

Quinlan [92] acknowledges the fact that the information gain is biased towards attributes with many values. This is a consequence of the previously mentioned particularity regarding entropy, in which further refinement leads to a decrease in its value. Quinlan proposes a solution for this matter called *gain ratio* [89]. It basically consists of normalizing the information gain by the entropy of the attribute being tested, that is,

$$\Delta\Phi^{gainRatio}(a_i, \mathbf{X}, y) = \frac{\Delta\Phi^{IG}}{\phi^{entropy}(a_i, \mathbf{X})}. \tag{2.5}$$

The gain ratio compensates the decrease in entropy in multiple partitions by dividing the information gain by the attribute self-entropy $\phi^{entropy}(a_i, \mathbf{X})$. The value of $\phi^{entropy}(a_i, \mathbf{X})$ increases logarithmically as the number of partitions over a_i increases, decreasing the value of gain ratio. Nevertheless, the gain ratio has two deficiencies: (i) it may be undefined (i.e., the value of self-entropy may be zero); and (ii) it may choose attributes with very low self-entropy but not with high gain. For solving these issues, Quinlan suggests first calculating the information gain for all attributes, and then calculating the gain ratio only for those cases in which the information gain value is above the average value of all attributes.

Several variations of the gain ratio have been proposed. For instance, the *normalized gain* [52] replaces the denominator of gain ratio by $\log_2 |a_i|$. The authors demonstrate two theorems with cases in which the normalized gain works better than or at least equally as either information gain or gain ratio does. In the first theorem, they prove that if two attributes a_i and a_j partition the instance space in pure sub-partitions, and that if $|a_i| > |a_j|$, normalized gain will always prefer a_j over a_i, whereas gain ratio is dependent of the self-entropy values of a_i and a_j (which means gain ratio may choose the attribute that partitions the space in more values). The second theorem states that given two attributes a_i and a_j, $|a_i| = |a_j|$, $|a_i| \geq 2$, if a_i partitions the instance space in pure subsets and a_j has at least one subset with more than one class, normalized gain will always prefer a_i over a_j, whereas gain ratio will prefer a_j if the following condition is met:

$$\frac{E(a_j, \mathbf{X}, y)}{\phi^{entropy}(y, \mathbf{X})} \leq 1 - \frac{\phi^{entropy}(a_j, \mathbf{X})}{\phi^{entropy}(a_i, \mathbf{X})}$$

where:

$$E(a_j, \mathbf{X}, y) = -\sum_{l=1}^{|a_j|} p_{v_l, \bullet} \times \phi^{entropy}(y, \mathbf{X}_{\mathbf{a_j}=\mathbf{v_l}}) \tag{2.6}$$

For details on the proof of each theorem, please refer to Jun et al. [52].

Other variation is the *average gain* [123], that replaces the denominator of gain ratio by $|dom(a_i)|$ (it only works for nominal attributes). The authors do not demonstrate theoretically any situations in which this measure is a better option than gain ratio. Their work is supported by empirical experiments in which the average gain outperforms gain ratio in terms of runtime and tree size, though with no significant differences regarding accuracy. Note that most decision-tree induction algorithms provide one branch for each nominal value an attribute can take. Hence, the average gain [123] is practically identical to the normalized gain [52], though without scaling the number of values with log_2.

Sá et al. [100] propose a somewhat different splitting measure based on the *minimum entropy of error* principle (MEE) [106]. It does not directly depend on the class distribution of a node p_{v_j,y_l} and the prevalences $p_{v_j,\bullet}$, but instead it depends on the errors produced by the decision rule on the form of a Stoller split [28]: if $a_i(x) \leq \Delta$, $y(x) = y_\omega$; $\hat{y}$ otherwise. In a Stoller split, each node split is binary and has an associated class y_ω for the case $a_i(x) \leq \Delta$, while the remaining classes are denoted by $\hat{y}$ and associated to the complementary branch. Each class is assigned a code $t \in \{-1, 1\}$, in such a way that for $y(x) = y_\omega$, $t = 1$ and for $y(x) = \hat{y}$, $t = -1$. The splitting measure is thus given by:

$$MEE(a_i, \mathbf{X}, y) = -(P_{-1}log_e P_{-1} + P_0 log_e P_0 + P_1 log_e P_1)$$

where:

$$P_{-1} = \frac{N_{\bullet,y_l}}{n} \times \frac{e_{1,-1}}{N_x}$$

$$P_1 = \left(1 - \frac{N_{\bullet,y_l}}{n}\right) \times \frac{e_{-1,1}}{N_x}$$

$$P_0 = 1 - P_{-1} - P_1 \tag{2.7}$$

where $e_{t,t'}$ is the number of instances t classified as t'. Note that unlike other measures such as information gain and gain ratio, there is no need of computing the impurity of sub-partitions and their subsequent average, as MEE does all the calculation needed at the current node to be split. MEE is bounded by the interval $[0, log_e 3]$, and needs to be minimized. The meaning of minimizing MEE is constraining the probability mass function of the errors to be as narrow as possible (around zero). The authors argue that by using MEE, there is no need of applying the pruning operation, saving execution time of decision-tree induction algorithms.

2.3.1.2 Distance-Based Criteria

Criteria in this category evaluate separability, divergency or discrimination between classes. They measure the distance between class probability distributions.

A popular distance criterion which is also from the class of impurity-based criteria is the *Gini index* [12, 39, 88]. It is given by:

$$\phi^{Gini}(y, \mathbf{X}) = 1 - \sum_{l=1}^{k} p_{\bullet,y_l}^2 \tag{2.8}$$

Breiman et al. [12] also acknowledge Gini's bias towards attributes with many values. They propose the *twoing* binary criterion for solving this matter. It belongs to the class of binary criteria, which requires attributes to have their domain split into two mutually exclusive subdomains, allowing binary splits only. For every binary criteria, the process of dividing attribute a_i values into two subdomains, d_1 and d_2, is exhaustive[1] and the division that maximizes its value is selected for attribute a_i. In other words, a binary criterion β is tested over all possible subdomains in order to provide the optimal binary split, β^*:

$$\beta^* = \max_{d_1, d_2} \beta(a_i, d_1, d_2, \mathbf{X}, y)$$

$$s.t.$$

$$d_1 \cup d_2 = dom(a_i)$$

$$d_1 \cap d_2 = \emptyset \tag{2.9}$$

Now that we have defined binary criteria, the twoing binary criterion is given by:

$$\beta^{twoing}(a_i, d_1, d_2, \mathbf{X}, y) = 0.25 \times p_{d_1, \bullet} \times p_{d_2, \bullet} \times \left(\sum_{l=1}^{k} abs(p_{y_l | d_1} - p_{y_l | d_2}) \right)^2 \tag{2.10}$$

where $abs(.)$ returns the absolute value.

Friedman [38] and Rounds [99] propose a binary criterion based on the Kolmogorov-Smirnoff (KS) distance for handling binary-class problems:

$$\beta^{KS}(a_i, d_1, d_2, \mathbf{X}, y) = abs(p_{d_1 | y_1} - p_{d_1 | y_2}) \tag{2.11}$$

Haskell and Noui-Mehidi [45] propose extending β^{KS} for handling multi-class problems. Utgoff and Clouse [120] also propose a multi-class extension to β^{KS}, as well as missing data treatment, and they present empirical results which show their criterion is similar in accuracy to Quinlan's gain ratio, but produces smaller-sized trees.

The χ^2 statistic [72, 125, 130] has been employed as a splitting criterion in decision trees. It compares the observed values with those that one would expect if there were no association between attribute and class. The resulting statistic is distributed approximately as the chi-square distribution, with larger values indicating greater association. Since we are looking for the predictive attribute with the highest degree of association to the class attribute, this measure must be maximized. It can be calculated as:

---

[1] Coppersmith et al. [25] present an interesting heuristic procedure for finding subdomains d_1 and d_2 when the partition is based on a nominal attribute. Shih [105] also investigates the problem of efficiency on finding the best binary split for nominal attributes.

$$\chi^2(a_i, \mathbf{X}, y) = \sum_{j=1}^{|a_i|} \sum_{l=1}^{k} \frac{\left(N_{v_j \cap y_l} - \frac{N_{v_j, \bullet} \times N_{\bullet, y_l}}{N_x}\right)^2}{\frac{N_{v_j, \bullet} \times N_{\bullet, y_l}}{N_x}} \tag{2.12}$$

It should be noticed that the χ^2 statistic (and similarly the G statistic) become poor approximations with small expected frequencies. Small frequencies make χ^2 over-optimistic in detecting informative attributes, i.e., the probability derived from the distribution will be smaller than the true probability of getting a value of χ^2 as large as that obtained.

De Mántaras [26] proposes a distance criterion that "provides a clearer and more formal framework for attribute selection and solves the problem of bias in favor of multivalued attributes without having the limitations of Quinlan's Gain Ratio". It is actually the same normalization to information gain as CAIR is to GMI, i.e.,

$$1 - \Delta\Phi^{distance}(a_i, \mathbf{X}, y) = \frac{\Delta\Phi^{IG}}{-\sum_{j=1}^{|a_i|} \sum_{l=1}^{k} p_{v_j \cap y_l} \log_2 p_{v_j \cap y_l}}. \tag{2.13}$$

Notice that in (2.13), we are actually presenting the complement of Mántaras distance measure, i.e., $1 - \Delta\Phi^{distance}(a_i, \mathbf{X}, y)$, but with no implications in the final result (apart from the fact that (2.13) needs to be maximized, whereas the original distance measure should be minimized).

Fayyad and Irani [35] propose a new family of measures called C-SEP (from Class SEParation). They claim that splitting measures such as information gain (and similar impurity-based criteria) suffer from a series of deficiencies (e.g., they are insensitive to within-class fragmentation), and they present new requirements a "good" splitting measure $\Gamma(.)$ (in particular, binary criteria) should fulfill:

- $\Gamma(.)$ is maximum when classes in d_1 and d_2 are disjoint (inter-class separability);
- $\Gamma(.)$ is minimum when the class distributions in d_1 and d_2 are identical;
- $\Gamma(.)$ favors partitions which keep instances from the same class in the same sub-domain d_i (intra-class cohesiveness);
- $\Gamma(.)$ is sensitive to permutations in the class distribution;
- $\Gamma(.)$ is non-negative, smooth (differentiable), and symmetric with respect to the classes.

Binary criteria that fulfill the above requirements are based on the premise that a good split is the one that separates as many different classes from each other as possible, while keeping examples of the same class together. Γ must be maximized, unlike the previously presented impurity-based criteria.

Fayyad and Irani [35] propose a new binary criterion from this family of C-SEP measures called *ORT*, defined as:

$$\Gamma^{ORT}(a_i, d_1, d_2, \mathbf{X}, y) = 1 - \theta(v_{d_1}, v_{d_2})$$

$$\theta(v_{d_1}, v_{d_2}) = \frac{v_{d_1} \cdot v_{d_2}}{||v_{d_1}|| \times ||v_{d_2}||} \tag{2.14}$$

where v_{d_i} is the class vector of the set of instances $\mathbf{X_i} = \{x \in \mathbf{X} \mid \mathbf{X_{a_i \in d_i}}\}$, "·" represents the inner product between two vectors and $\|.\|$ the magnitude (norm) of a vector. Note that *ORT* is basically the complement of the well-known *cosine* distance, which measures the orthogonality between two vectors. When the angle between two vectors is 90, it means the non-zero components of each vector do not overlap. The *ORT* criterion is maximum when the cosine distance is minimum, i.e., the vectors are orthogonal, and it is minimum when they are parallel. The higher the values of *ORT*, the greater the distance between components of the class vectors (maximum *ORT* means disjoint classes).

Taylor and Silverman [111] propose a splitting criterion called *mean posterior improvement* (MPI), which is given by:

$$\beta^{MPI}(a_i, d_1, d_2, \mathbf{X}, y) = p_{d_1, \bullet} p_{d_2, \bullet} - \sum_{l=1}^{k}[p_{\bullet, y_l} p_{d_1 \cap y_l} p_{d_2 \cap y_l}] \qquad (2.15)$$

The MPI criterion provides maximum value when individuals of the same class are all placed in the same partition, and thus, (2.15) should be maximized. Classes over-represented in the father node will have a greater emphasis in the MPI calculation (such an emphasis is given by the $p_{\bullet, y_l}$ in the summation). The term $p_{d_1 \cap y_l} p_{d_2 \cap y_l}$ is desired to be small since the goal of MPI is to keep instances of the same class together and to separate them from those of other classes. Hence, $p_{d_1, \bullet} p_{d_2, \bullet} - p_{d_1 \cap y_l} p_{d_2 \cap y_l}$ is the improvement that the split is making for class y_l, and therefore the MPI criterion is the mean improvement over all the classes.

Mola and Siciliano [75] propose using the predictability index τ originally proposed in [42] as a splitting measure. The τ index can be used first to evaluate each attribute individually (2.16), and then to evaluate each possible binary split provided by grouping the values of a given attribute in d_1 and d_2 (2.17).

$$\beta^{\tau}(a_i, \mathbf{X}, y) = \frac{\sum_{j=1}^{|a_i|} \sum_{l=1}^{k} (p_{v_j \cap y_l})^2 \times p_{v_j, \bullet} - \sum_{l=1}^{k} p_{\bullet, y_l}^2}{1 - \sum_{l=1}^{k} p_{\bullet, y_l}^2}$$

$$(2.16)$$

$$\beta^{\tau}(a_i, d_1, d_2, \mathbf{X}, y) = \frac{\sum_{l=1}^{k} (p_{d_1 \cap y_l})^2 p_{d_1, \bullet} + \sum_{l=1}^{k} (p_{d_2 \cap y_l})^2 p_{d_2, \bullet} - \sum_{l=1}^{k} p_{\bullet, y_l}^2}{1 - \sum_{j=1}^{k} p_{y_j}^2}$$

$$(2.17)$$

Now, consider that $\beta^{\tau*}(a_i) = \max_{d_1, d_2} \beta^{\tau}(a_i, d_1, d_2, \mathbf{X}, y)$. Mola and Siciliano [75] prove a theorem saying that

$$\beta^{\tau}(a_i, \mathbf{X}, y) \geq \beta^{\tau}(a_i, d_1, d_2, \mathbf{X}, y) \qquad (2.18)$$

and also that

$$\beta^{\tau}(a_i, \mathbf{X}, y) \geq \beta^{\tau*}(a_i). \qquad (2.19)$$

This theoretical evidence is of great importance for providing a means to select the best attribute and its corresponding binary partitions without the need of exhaustively trying all possibilities. More specifically, one has to calculate (2.16) for all attributes, and to sort them according to the highest values of $\beta^{\tau}(*, \mathbf{X}, y)$, in such a way that a_1 is the attribute that yields the highest value of $\beta^{\tau}(*, \mathbf{X}, y)$, a_2 is the second highest value, and so on. Then, one has to test all possible splitting options in (2.17) in order to find $\beta^{\tau*}(a_1)$. If the value of $\beta^{\tau*}(a_1)$ is greater than the value of $\beta^{\tau}(a_2, \mathbf{X}, y)$, we do not need to try any other split possibilities, since we know that $\beta^{\tau*}(a_2)$ is necessarily lesser than $\beta^{\tau}(a_2, \mathbf{X}, y)$. For a simple but efficient algorithm implementing this idea, please refer to the appendix in [75].

2.3.1.3 Other Classification Criteria

In this category, we include all criteria that did not fit in the previously-mentioned categories.

Li and Dubes [62] propose a binary criterion for binary-class problems called *permutation statistic*. It evaluates the degree of similarity between two vectors, V_{a_i} and y, and the larger this statistic, the more alike the vectors. Vector V_{a_i} is calculated as follows. Let a_i be a given numeric attribute with the values $[8.20, 7.3, 9.35, 4.8, 7.65, 4.33]$ and $N_x = 6$. Vector $y = [0, 0, 1, 1, 0, 1]$ holds the corresponding class labels. Now consider a given threshold $\Delta = 5.0$. Vector V_{a_i} is calculated in two steps: first, attribute a_i values are sorted, i.e., $a_i = [4.33, 4.8, 7.3, 7.65, 8.20, 9.35]$, consequently rearranging $y = [1, 1, 0, 0, 0, 1]$; then, $V_{a_i}(n)$ takes 0 when $a_i(n) \leq \Delta$, and 1 otherwise. Thus, $V_{a_i} = [0, 0, 1, 1, 1, 1]$. The permutation statistic first analyses how many $1-1$ matches (d) vectors V_{a_i} and y have. In this particular example, $d = 1$. Next, it counts how many 1's there are in $V_{a_i}(n_a)$ and in $y(n_y)$. Finally, the permutation statistic can be computed as:

$$\beta^{permutation}(V_{a_i}, y) = \sum_{j=0}^{d} \frac{\binom{n_a}{j}\binom{N_x-n_a}{n_y-j}}{\binom{N_x}{n_y}} - \frac{\binom{n_a}{d}\binom{N_x-n_a}{n_y-d}}{\binom{N_x}{n_y}} U$$

$$\binom{n}{m} = 0 \quad \text{if} \quad n < 0 \text{ or } m < 0 \text{ or } n < m$$

$$= \frac{n!}{m!(n-m)!} \text{otherwise} \qquad (2.20)$$

where U is a (continuous) random variable distributed uniformly over $[0, 1]$.

The permutation statistic presents an advantage over the information gain and other criteria: it is not sensitive to the data fragmentation problem.[2] It automatically adjusts for variations in the number of instances from node to node because its distribution does not change with the number of instances at each node.

Quinlan and Rivest [96] propose using the *minimum description length principle* (MDL) as a splitting measure for decision-tree induction. MDL states that, given a set of competing hypotheses (in this case, decision trees), one should choose as the preferred hypothesis the one that minimizes the sum of two terms: (i) the description length of the hypothesis (d_l); and (ii) length of the data given the hypothesis (l_h). In the context of decision trees, the second term can be regarded as *the length of the exceptions*, i.e., the length of certain objects of a given subset whose class value is different from the most frequent one. Both terms are measured in bits, and thus one needs to encode the decision tree and exceptions accordingly.

It can be noticed that by maximizing d_l, we minimize l_h, and vice-versa. For instance, when we grow a decision tree until each node has objects that belong to the same class, we usually end up with a large tree (maximum d_l) prone to overfitting, but with no exceptions (minimum l_h). Conversely, if we allow a large number of exceptions, we will not need to partition subsets any further, and in the extreme case (maximum l_h), the decision tree will hold a single leaf node labeled as the most frequent class value (minimum d_l). Hence the need of minimizing the sum $d_l + l_h$.

MDL provides a way of comparing decision trees once the encoding techniques are chosen. Finding a suitable encoding scheme is usually a very difficult task, and the values of d_l and l_h are quite dependent on the encoding technique used [37]. Nevertheless, Quinlan and Rivest [96] propose selecting the attribute that minimizes $d_l + l_h$ at each node, and then pruning back the tree whenever replacing an internal node by a leaf decreases $d_l + l_h$.

A criterion derived from classical statistics is the *multiple hypergeometric distribution* (P_0) [1, 70], which is an extension of Fischer's *exact test* for two binary variables. It can be regarded as the probability of obtaining the observed data given that the null hypothesis (of variable independence) is true. P_0 is given by:

$$P_0(a_i, \mathbf{X}, y) = \left(\frac{\prod_{l=1}^{k} N_{\bullet, y_l}!}{N_x!} \right) \prod_{j=1}^{|a_i|} \left(\frac{N_{v_j, \bullet}!}{\prod_{m=1}^{k} N_{v_j \cap y_m}!} \right) \tag{2.21}$$

The lower the values of P_0, the lower the probability of accepting the null hypothesis. Hence, the attribute that presents the lowest value of P_0 is chosen for splitting the current node in a decision tree.

Chandra and Varghese [21] propose a new splitting criterion for partitioning nodes in decision trees. The proposed measure is designed to reduce the number of distinct classes resulting in each sub-tree after a split. Since the authors do not name

---

[2] Data fragmentation is a well-known problem in top-down decision trees. Nodes with few instances usually lack statistical support for further partitioning. This phenomenon happens for most of the split criteria available, since their distributions depend on the number of instances in each particular node.

their proposed measure, we call it CV (from Chandra-Varseghe) from now on. It is given by:

$$CV(a_i, \mathbf{X}, y) = \sum_{j=1}^{|a_i|} \left[p_{v_j,\bullet} \times \frac{D_{v_j}}{D_x} \left(\sum_{l=1}^{D_{v_j}} p_{v_j|y_l} \right) \right] \qquad (2.22)$$

where D_x counts the number of *distinct* class values among the set of instances in $\mathbf{X}$, and D_{v_j} the number of distinct class values in partition v_j. The CV criterion must be minimized. The authors prove that CV is strictly convex (i.e., it achieves its minimum value at a boundary point) and cumulative (and thus, well-behaved). The authors argue that the experiments, which were performed on 19 data sets from the UCI repository [36], indicate that the proposed measure results in decision trees that are more compact (in terms of tree height), without compromising on accuracy when compared to the gain ratio and Gini index.

Chandra et al. [20] propose the use of a *distinct class based splitting measure* (DCSM). It is given by:

$$DCSM(a_i, \mathbf{X}, y) = \sum_{j=1}^{|a_i|} \left[p_{v_j,\bullet} D_{v_j} \exp(D_{v_j}) \right.$$

$$\left. \times \sum_{l=1}^{k} \left[p_{y_l|v_j} \exp \left(\frac{D_{v_j}}{D_x} \left(1 - \left(p_{y_l|v_j} \right)^2 \right) \right) \right] \right] \qquad (2.23)$$

Note that the term $D_{v_j} \exp D_{v_j}$ deals with the number of distinct classes in a given partition v_j. As the number of distinct classes in a given partition increases, this term also increases. It means that purer partitions are preferred, and they are weighted according to the proportion of training instances that lie in the given partition. Also, note that $\frac{D_{v_j}}{D_x}$ decreases when the number of distinct classes decreases while $(1 - (p_{y_l|v_j})^2)$ decreases when there are more instances of a class compared to the total number of instances in a partition. These terms also favor partitions with a small number of distinct classes.

It can be noticed that the value of DCSM increases exponentially as the number of distinct classes in the partition increases, invalidating such splits. Chandra et al. [20] argue that "this makes the measure more sensitive to the impurities present in the partition as compared to the existing measures." The authors demonstrate that DCSM satisfies two properties: convexity and well-behavedness. Finally, through empirical data, the authors affirm that DCSM provides more compact and more accurate trees than those provided by measures such as gain ratio and Gini index.

Many other split criteria for classification can be found in the literature, including *relevance* [3], *misclassification error with confidence intervals* [53], *RELIEF split criterion* [58], *QUEST split criterion* [65], just to name a few.

2.3.1.4 Regression Criteria

All criteria presented so far are dedicated to classification problems. For regression problems, where the target variable y is continuous, a common approach is to calculate the *mean squared error* (MSE) as a splitting criterion:

$$MSE(a_i, \mathbf{X}, y) = N_x^{-1} \sum_{j=1}^{|a_i|} \sum_{x_l \in v_j} (y(x_l) - \bar{y}_{v_j})^2 \tag{2.24}$$

where $\bar{y}_{v_j} = N_{v_{j,\bullet}}^{-1} \sum_{x_l \in v_j} y(x_l)$. Just as with clustering, we are trying to minimize the within-partition variance. Usually, the sum of squared errors is weighted over each partition according to the estimated probability of an instance belonging to the given partition [12]. Thus, we should rewrite MSE to:

$$wMSE(a_i, \mathbf{X}, y) = \sum_{j=1}^{|a_i|} p_{v_{j,\bullet}} \sum_{x_l \in v_j} (y(x_l) - \bar{y}_{v_j})^2 \tag{2.25}$$

Another common criterion for regression is the *sum of absolute deviations* (SAD) [12], or similarly its weighted version given by:

$$wSAD(a_i, \mathbf{X}, y) = \sum_{j=1}^{|a_i|} p_{v_{j,\bullet}} \sum_{x_l \in v_j} abs(y(x_l) - median(y_{v_j})) \tag{2.26}$$

where $median(y_{v_j})$ is the target attribute's median of instances belonging to $\mathbf{X}_{a_i=v_j}$.

Quinlan [93] proposes the use of the *standard deviation reduction* (SDR) for his pioneering system of model trees induction, M5. Wang and Witten [124] extend the work of Quinlan in their proposed system M5', also employing the SDR criterion. It is given by:

$$SDR(a_i, \mathbf{X}, y) = \sigma_X - \sum_{j=1}^{|a_i|} p_{v_{j,\bullet}} \sigma_{v_j} \tag{2.27}$$

where σ_X is the standard deviation of instances in $\mathbf{X}$ and σ_{v_j} the standard deviation of instances in $\mathbf{X}_{a_i=v_j}$. SDR should be maximized, i.e., the weighted sum of standard deviations of each partition should be small as possible. Thus, partitioning the instance space according to a particular attribute a_i should provide partitions whose target attribute variance is small (once again we are interested in minimizing the within-partition variance). Observe that minimizing the second term in SDR is equivalent to minimizing wMSE, but in SDR we are using the partition standard deviation (σ) as a similarity criterion whereas in wMSE we are using the partition variance (σ^2).

Buja and Lee [15] propose two alternative regression criteria for binary trees: *one-sided purity* (OSP) and *one-sided extremes* (OSE). OSP is defined as:

$$OSP(a_i, d_1, d_2, \mathbf{X}, y) = \min_{d_1, d_2}(\sigma^2_{d_1}, \sigma^2_{d_2}) \tag{2.28}$$

where $\sigma^2_{d_i}$ is the variance of partition d_i. The authors argue that by minimizing this criterion for all possible splits, we find a split whose partition (either d_1 or d_2) presents the smallest variance. Typically, this partition is less likely to be split again. Buja and Lee [15] also propose the OSE criterion:

$$OSE(a_i, d_1, d_2, \mathbf{X}, y) = \min_{d_1, d_2}(\bar{y_{d_1}}, \bar{y_{d_2}})$$

or, conservely:

$$OSE(a_i, d_1, d_2, \mathbf{X}, y) = \max_{d_1, d_2}(\bar{y_{d_1}}, \bar{y_{d_2}}) \tag{2.29}$$

The authors argue that whereas the mean values have not been thought of as splitting criteria, "in real data, the dependence of the mean response on the predictor variables is often monotone; hence extreme response values are often found on the periphery of variable ranges (…), the kind of situations to each the OSE criteria would respond".

Alpaydin [2] mentions the use of the *worst possible error* (WPE) as a valid criterion for splitting nodes:

$$WPE(a_i, \mathbf{X}, y) = \max_{j} \max_{l}[abs(y(x_l) - \bar{y_{v_j}})] \tag{2.30}$$

Alpaydin [2] states that by using WPE we can guarantee that the error for any instance is never larger than a given threshold Δ. This analysis is useful because the threshold Δ can be seen as a complexity parameter that defines the *fitting level* provided by the tree, given that we use it for deciding when interrupting its growth. Larger values of Δ lead to smaller trees that could underfit the data whereas smaller values of Δ lead to larger trees that risk overfitting. A deeper appreciation of underfitting, overfitting and tree complexity is presented later, when *pruning* is discussed.

Other regression criteria that can be found in the literature are MPI for regression [112], Lee's criteria [61], GUIDE's criterion [66], and SMOTI's criterion [67], just to name a few. Tables 2.1 and 2.2 show all univariate splitting criteria cited in this section, as well as their corresponding references, listed in chronological order.

2.3.1.5 Multivariate Splits

All criteria presented so far are intended for building univariate splits. Decision trees with multivariate splits (known as *oblique*, *linear* or *multivariate* decision trees) are not so popular as the univariate ones, mainly because they are harder to interpret. Nevertheless, researchers reckon that multivariate splits can improve the performance

Table 2.1 Univariate splitting criteria for classification

Category	Criterion	References
Info theory	GMI	[22, 41, 102, 108]
	Information gain	[18, 44, 92, 122]
	G statistic	[72, 73]
	Gain ratio	[89, 92]
	CAIR	[22]
	Normalized gain	[52]
	Average gain	[123]
	MEE	[100]
Distance-based	KS distance	[38, 99]
	Gini index	[12]
	Twoing	[12]
	χ^2	[72, 125, 130]
	Distance	[26]
	Multi-class KS	[45, 120]
	ORT	[35]
	MPI	[111]
	τ Index	[75]
Other	Permutation	[62]
	Relevance	[3]
	MDL criterion	[96]
	Mis. Error with CI	[53]
	RELIEF	[58]
	QUEST criterion	[65]
	P_0	[1, 70]
	CV	[21]
	DCSM	[20]

Table 2.2 Univariate splitting criteria for regression

Criterion	References
(w)MSE	[12]
(w)SAD	[12]
SDR	[93, 124]
MPI-R	[112]
OSP	[15]
OSE	[15]
Lee's	[61]
GUIDE's	[66]
SMOTI's	[67]
WPE	[2]

of the tree in several data sets, while generating smaller trees [47, 77, 98]. Clearly, there is a tradeoff to consider in allowing multivariate tests: simple tests may result in large trees that are hard to understand, yet multivariate tests may result in small trees with tests hard to understand [121].

A decision tree with multivariate splits is able to produce polygonal (polyhedral) partitions of the attribute space (hyperplanes at an oblique orientation to the attribute axes) whereas univariate trees can only produce hyper-rectangles parallel to the attribute axes. The tests at each node have the form:

$$w_0 + \sum_{i=1}^{n} w_i a_i(x) \le 0 \qquad (2.31)$$

where w_i is a real-valued coefficient associated to the ith attribute and w_0 the disturbance coefficient of the test.

CART (Classification and Regression Trees) [12] is one of the first systems that allowed multivariate splits. It employs a hill-climbing strategy with a backward attribute elimination for finding good (albeit suboptimal) linear combinations of attributes in non-terminal nodes. It is a fully-deterministic algorithm with no built-in mechanisms to escape local-optima. Breiman et al. [12] point out that the proposed algorithm has much room for improvement.

Another approach for building oblique decision trees is LMDT (Linear Machine Decision Trees) [14, 119], which is an evolution of the perceptron tree method [117]. Each non-terminal node holds a linear machine [83], which is a set of k linear discriminant functions that are used collectively to assign an instance to one of the k existing classes. LMDT uses heuristics to determine when a linear machine has stabilized (since convergence cannot be guaranteed). More specifically, for handling non-linearly separable problems, a method similar to simulated annealing (SA) is used (called *thermal training*). Draper and Brodley [30] show how LMDT can be altered to induce decision trees that minimize arbitrary misclassification cost functions.

SADT (Simulated Annealing of Decision Trees) [47] is a system that employs SA for finding good coefficient values for attributes in non-terminal nodes of decision trees. First, it places a hyperplane in a canonical location, and then iteratively perturbs the coefficients in small random amounts. At the beginning, when the temperature parameter of the SA is high, practically any perturbation of the coefficients is accepted regardless of the goodness-of-split value (the value of the utilised splitting criterion). As the SA cools down, only perturbations that improve the goodness-of-split are likely to be allowed. Although SADT can eventually escape from local-optima, its efficiency is compromised since it may consider tens of thousands of hyperplanes in a single node during annealing.

OC1 (Oblique Classifier 1) [77, 80] is yet another oblique decision tree system. It is a thorough extension of CART's oblique decision tree strategy. OC1 presents the advantage of being more efficient than the previously described systems. For instance, in the worst case scenario, OC1's running time is $O(\log_n)$ times greater than the worst case scenario of univariate decision trees, i.e., $O(nN^2 \log_N)$ *versus* $O(nN^2)$. OC1 searches for the best univariate split as well as the best oblique split, and it only employs the oblique split when it improves over the univariate split.[3] It uses both a deterministic heuristic search (as employed in CART) for finding local-optima and a non-deterministic search (as employed in SADT—though not SA) for escaping local-optima.

During the deterministic search, OC1 perturbs the hyperplane coefficients sequentially (much in the same way CART does) until no significant gain is achieved according to an impurity measure. More specifically, consider hyperplane $H = w_0 + \sum_{i=1}^{n} w_i a_i(x) = 0$, and that we substitute an instance x_j in H, i.e., $H = w_0 + \sum_{i=1}^{n} w_i a_i(x_j) = Z_j$. The sign of Z_j indicates whether an instance x_j is above or below the hyperplane H. If H splits $\mathbf{X}$ perfectly, then all instances belonging to the same class will have the same sign of Z. For finding the local-optimal set of coefficients, OC1 employs a sequential procedure that works as follows: treat coefficient w_i as a variable and all other coefficients as constants. The condition that instance x_j is above hyperplane H can be written as:

$$Z_j > 0$$

$$w_i > \left[\frac{w_i a_i(x_j) - Z_j}{a_i(x_j)} \equiv U_j \right] \tag{2.32}$$

assuming $a_i(x_j) > 0$, which is ensured through normalization. With the definition in (2.32), an instance is above the hyperplane if $w_i > U_j$ and below otherwise. By plugging each instance $x \in \mathbf{X}$ in (2.32), we obtain N_x constraints on the value of w_i. Hence, the problem is reduced on finding the value of w_i that satisfies the greatest possible number of constraints. This problem is easy to solve optimally: simply sort all the values U_j, and consider setting w_i to the midpoint between each pair of different class values. For each distinct placement of the coefficient w_i, OC1 computes the impurity of the resulting split, and replaces original coefficient w_i by the recently discovered value if there is reduction on impurity. The pseudocode of this deterministic perturbation method is presented in Algorithm 2.

The parameter P_{stag} (stagnation probability) is the probability that a hyperplane is perturbed to a location that does not change the impurity measure. To prevent the stagnation of impurity, P_{stag} decreases by a constant amount each time OC1 makes a "stagnant" perturbation, which means only a constant number of such perturbations will occur at each node. P_{stag} is reset to 1 every time the global impurity measure is improved. It is a user-defined parameter.

After a local-optimal hyperplane H is found, it is further perturbed by a randomized vector, as follows: it computes the optimal amount by which H should be perturbed along the random direction dictated by a random vector. To be more precise, when a hyperplane $H = w_0 + \sum_{i=1}^{n} w_i a_i(x)$ cannot be improved by

---

[3] OC1 only allows the option of employing oblique splits when $N > 2n$, though this threshold can be user-defined.

Algorithm 2 Deterministic OC1's procedure for perturbing a given coefficient. Parameters are the current hyperplane H and the coefficient index i.

```
1: procedure PERTURB(H, i)
2:     for j = 1 to N_x do
3:         Compute U_j (32)
4:     end for
5:     Sort U_1..U_{N_x} in non-decreasing order
6:     w'_i = best split of the sorted U_j s
7:     H_1 = resulting of replacing w_i by w'_i in H
8:     if (impurity(H_1) < impurity(H)) then
9:         w_i = w'_i
10:        P_move = P_stag
11:    else if (impurity(H_1) = impurity(H)) then
12:        w_i = w'_i with probability P_move
13:        P_move = P_move − 0.1 P_stag
14:    end ifreturn w_i
15: end procedure
```

deterministic perturbation (Algorithm 2), OC1 repeats the following loop J times (where J is a user-specified parameter, set to 5 by default):

- Choose a random vector $R = [r_0, r_1, \ldots, r_n]$;
- Let α be the amount by which we want to perturb H in the direction R. More specifically, let $H_1 = (w_0 + \alpha r_0) + \sum_{i=1}^{n} (w_i + \alpha r_i) a_i(x)$;
- Find the optimal value for α;
- If the hyperplane H_1 decreases the overall impurity, replace H with $H1$, exit this loop and begin the deterministic perturbation algorithm for the individual coefficients.

Note that we can treat α as the only variable in the equation for H_1. Therefore each of the N examples, if plugged into the equation for H_1, imposes a constraint on the value of α. OC1 can use its own deterministic coefficient perturbation method (Algorithm 2) to compute the best value of α. If J *random jumps* fail to improve the impurity measure, OC1 halts and uses H as the split for the current tree node. Regarding the impurity measure, OC1 allows the user to choose among a set of splitting criteria, such as information gain, Gini index, twoing criterion, among others.

Ittner [51] proposes using OC1 over an augmented attribute space, generating *nonlinear decision trees*. The key idea involved is to "build" new attributes by considering all possible pairwise products and squares of the original set of n attributes. As a result, a new attribute space with $(n^2 + 3n)/2$ is formed, i.e., the sum of n original attributes, n squared ones and $(n(n-1))/2$ pairwise products of the original attributes. To illustrate, consider a binary attribute space $\{a_1, a_2\}$. The augmented attribute space would contain 5 attributes, i.e., $b_1 = a_1, b_2 = a_2, b_3 = a_1 a_2, b_4 = a_1^2, b_5 = a_2^2$.

A similar approach of transforming the original attributes is taken in [64], in which the authors propose the BMDT system. In BMDT, a 2-layer feedforward neural network is employed to transform the original attribute space in a space in which the new attributes are linear combinations of the original ones. This transformation is performed through a hyperbolic tangent function at the hidden units. After transforming the attributes, a univariate decision-tree induction algorithm is employed over this new attribute space. Finally, a procedure replaces the transformed attributes by the original ones, which means that the univariate tests in the recently built decision tree become multivariate tests, and thus the univariate tree becomes an oblique tree.

Shah and Sastry [103] propose the APDT (Alopex Perceptron Decision Tree) system. It is an oblique decision tree inducer that makes use of a new splitting criterion, based on the level of non-separability of the input instances. They argue that because oblique decision trees can realize arbitrary piecewise linear separating surfaces, it seems better to base the evaluation function on the degree of separability of the partitions rather than on the degree of purity of them. APDT runs the Perceptron algorithm for estimating the number of non-separable instances belonging to each one of the binary partitions provided by an initial hyperplane. Then, a correlation-based optimization algorithm called Unnikrishnan et al. [116] is employed for tuning the hyperplane weights taking into account the need of minimizing the new split criterion based on the degree of separability of partitions. Shah and Sastry [103] also propose a pruning algorithm based on genetic algorithms.

Several other oblique decision-tree systems were proposed employing different strategies for defining the weights of hyperplanes and for evaluating the generated split. Some examples include: the system proposed by Bobrowski and Kretowski [11], which employs heuristic sequential search (combination of sequential backward elimination and sequential forward selection) for defining hyperplanes and a dipolar criterion for evaluating splits; the omnivariate decision tree inducer proposed by Yildiz and Alpaydin [128], where the non-terminal nodes may be univariate, linear, or nonlinear depending on the outcome of comparative statistical tests on accuracy, allowing the split to match automatically the complexity of the node according to the subproblem defined by the data reaching that node; Li et al. [63] propose using tabu search and a variation of linear discriminant analysis for generating multivariate splits, arguing that their algorithm runs faster than most oblique tree inducers, since its computing time increases linearly with the number of instances; Tan and Dowe [109] proposes inducing oblique trees through a MDL-based splitting criterion and the evolution strategy as a meta-heuristic to search for the optimal hyperplane within a node. For regression oblique trees please refer to [27, 48, 60].

For the interested reader, it is worth mentioning that there are methods that induce oblique decision trees with optimal hyperplanes, discovered through *linear programming* [9, 10, 68]. Though these methods can find the optimal hyperplanes for specific splitting measures, the size of the linear program grows very fast with the number of instances and attributes.

For a discussion on several papers that employ evolutionary algorithms for induction of oblique decision trees (to evolve either the hyperplanes or the whole tree), the reader is referred to [**Barros2012**]. Table 2.3 presents a summary of some systems proposed for induction of oblique decision trees.

Table 2.3 Multivariate splits

System	Criterion	Hyperplane strategy
CART [12]	Gini index/twoing	Hill-climbing with SBE
LMDT [14]	Misclassification error	Linear machine with thermal training
SADT [47]	Sum-minority	Simulated annealing
OC1 [77, 80]	Info gain, gini index, twoing, etc	Hill-climbing with randomization
BMDT [64]	Gain ratio	2-Layer feedforward neural network
APDT [103]	Separability[a]	Alopex
Bobrowski and Kretowski [11][b]	Dipolar criterion	Heuristic sequential search
Omni [128]	Misclassification error	MLP neural network
LDTS [63]	Info gain	LDA with tabu search
MML [63]	MDL-based	Evolution strategy
Geometric DT [69]	Gini-index	Multisurface proximal SVM

[a]The authors of the criterion do not explicitly name it, so we call it "Separability", since the criterion is based on the degree of linear separability
[b]The authors call their system "our method", so we do not explicitly name it

2.3.2 Stopping Criteria

The top-down induction of a decision tree is recursive and it continues until a stopping criterion (or some stopping criteria) is satisfied. Some popular stopping criteria are [32, 98]:

1. Reaching class homogeneity: when all instances that reach a given node belong to the same class, there is no reason to split this node any further;
2. Reaching attribute homogeneity: when all instances that reach a given node have the same attribute values (though not necessarily the same class value);
3. Reaching the maximum tree depth: a parameter *tree depth* can be specified to avoid deep trees;
4. Reaching the minimum number of instances for a non-terminal node: a parameter *minimum number of instances for a non-terminal node* can be specified to avoid (or at least alleviate) the data fragmentation problem;
5. Failing to exceed a threshold when calculating the splitting criterion: a parameter *splitting criterion threshold* can be specified for avoiding *weak* splits.

Criterion 1 is universally accepted and it is implemented in most top-down decision-tree induction algorithms to date. Criterion 2 deals with the case of contradictory instances, i.e., identical instances regarding *A*, but with different class values. Criterion 3 is usually a constraint regarding tree complexity, specially for those cases in which comprehensibility is an important requirement, though it may affect complex classification problems which require deeper trees. Criterion 4 implies that small disjuncts (i.e., tree leaves covering a small number of objects) can be ignored since they are error-prone. Note that eliminating small disjuncts can be harmful to exceptions—particularly in a scenario of imbalanced classes. Criterion 5 is heavily dependent on the splitting measure used. An example presented in [32] clearly indicates a scenario in which using criterion 5 prevents the growth of a 100 % accurate decision tree (a problem usually referred to as *the horizon effect* [12, 89]).

The five criteria presented above can be seen as *pre-pruning* strategies, since they "prematurely" interrupt the growth of the tree. Note that most of the criteria discussed here may harm the growth of an accurate decision tree. Indeed there is practically a consensus in the literature that decision trees should be overgrown instead. For that, the stopping criterion used should be as *loose* as possible (e.g., until a single instance is contemplated by the node or until criterion 1 is satisfied). Then, a *post-pruning* technique should be employed in order to prevent *data overfitting*—a phenomenon that happens when the classifier *over-learns* the data, that is, when it learns all data peculiarities—including potential noise and spurious patterns—that are specific to the training set and do not generalise well to the test set. Post-pruning techniques are covered in the next section.

2.3.3 Pruning

This section reviews strategies of pruning, normally referred to as *post-pruning* techniques. Pruning is usually performed in decision trees for enhancing tree comprehensibility (by reducing its size) while maintaining (or even improving) accuracy. It was originally conceived as a strategy for tolerating noisy data, though it was found that it could improve decision tree accuracy in many noisy data sets [12, 92, 94].

A pruning method receives as input an unpruned tree T and outputs a decision tree T' formed by removing one or more subtrees from T. It replaces non-terminal nodes by leaf nodes according to a given heuristic. Next, we present the six most well-known pruning methods for decision trees [13, 32]: (1) reduced-error pruning; (2) pessimistic error pruning; (3) minimum error pruning; (4) critical-value pruning; (5) cost-complexity pruning; and (6) error-based pruning.

2.3.3.1 Reduced-Error Pruning

Reduced-error pruning is a conceptually simple strategy proposed by Quinlan [94]. It uses a pruning set (a part of the training set) to evaluate the goodness of a given subtree from T. The idea is to evaluate each non-terminal node $t \in \zeta_T$ with regard to the classification error in the pruning set. If such an error decreases when we replace the subtree $T^{(t)}$ by a leaf node, than $T^{(t)}$ must be pruned.

Quinlan imposes a constraint: a node t cannot be pruned if it contains a subtree that yields a lower classification error in the pruning set. The practical consequence of this constraint is that REP should be performed in a bottom-up fashion. The REP pruned tree T' presents an interesting optimality property: it is the smallest most accurate tree resulting from pruning original tree T [94]. Besides this optimality property, another advantage of REP is its linear complexity, since each node is visited only once in T. An obvious disadvantage is the need of using a pruning set, which means one has to divide the original training set, resulting in less instances to grow the tree. This disadvantage is particularly serious for small data sets.

2.3.3.2 Pessimistic Error Pruning

Also proposed by Quinlan [94], the pessimistic error pruning uses the training set for both growing and pruning the tree. The apparent error rate, i.e., the error rate calculated over the training set, is optimistically biased and cannot be used to decide whether pruning should be performed or not. Quinlan thus proposes adjusting the apparent error according to the continuity correction for the binomial distribution (cc) in order to provide a more realistic error rate. Consider the apparent error of a pruned node t, and the error of its entire subtree $T^{(t)}$ before pruning is performed, respectively:

$$r^{(t)} = \frac{E^{(t)}}{N_x^{(t)}} \tag{2.33}$$

$$r^{T^{(t)}} = \frac{\sum_{s \in \lambda_{T(t)}} E^{(s)}}{\sum_{s \in \lambda_{T(t)}} N_x^{(s)}}. \tag{2.34}$$

Modifying (2.33) and (2.34) according to cc results in:

$$r_{cc}^{(t)} = \frac{E^{(t)} + 1/2}{N_x^{(t)}} \tag{2.35}$$

$$r_{cc}^{T^{(t)}} = \frac{\sum_{s \in \lambda_{T(t)}} E^{(s)} + 1/2}{\sum_{s \in \lambda_{T(t)}} N_x^{(s)}} = \frac{\frac{|\lambda_{T(t)}|}{2} \sum_{s \in \lambda_{T(t)}} E^{(s)}}{\sum_{s \in \lambda_{T(t)}} N_x^{(s)}}. \tag{2.36}$$

For the sake of simplicity, we will refer to the adjusted number of errors rather than the adjusted error rate, i.e., $E_{cc}^{(t)} = E^{(t)} + 1/2$ and $E_{cc}^{T^{(t)}} = (|\lambda_{T(t)}|/2) \sum_{s \in \lambda_{T(t)}} E^{(s)}$. Ideally, pruning should occur if $E_{cc}^{(t)} \leq E_{cc}^{T^{(t)}}$, but note that this condition seldom holds, since the decision tree is usually grown up to the homogeneity stopping criterion (criterion 1 in Sect. 2.3.2), and thus $E_{cc}^{T^{(t)}} = |\lambda_{T(t)}|/2$ whereas $E_{cc}^{(t)}$ will very probably be a higher value. In fact, due to the homogeneity stopping criterion, $E_{cc}^{T^{(t)}}$ becomes simply a measure of complexity which associates each leaf node with a cost of $1/2$. Quinlan, aware of this situation, weakens the original condition

$$E_{cc}^{(t)} \leq E_{cc}^{T^{(t)}}$$ (2.37)

to

$$E_{cc}^{(t)} \leq E_{cc}^{T^{(t)}} + SE(E_{cc}^{T^{(t)}})$$ (2.38)

where

$$SE(E_{cc}^{T^{(t)}}) = \sqrt{\frac{E_{cc}^{T^{(t)}} * (N_x^{(t)} - E_{cc}^{T^{(t)}})}{N_x^{(t)}}}$$

is the standard error for the subtree $T^{(t)}$, computed as if the distribution of errors were binomial.

PEP is computed in a top-down fashion, and if a given node t is pruned, its descendants are not examined, which makes this pruning strategy quite efficient in terms of computational effort. As a point of criticism, Esposito et al. [32] point out that the introduction of the continuity correction in the estimation of the error rate has no theoretical justification, since it was never applied to correct over-optimistic estimates of error rates in statistics.

2.3.3.3 Minimum Error Pruning

Originally proposed by Niblett and Bratko [82] and further extended by Cestnik and Bartko [19], minimum error pruning is a bottom-up approach that seeks to minimize the *expected error rate* for unseen cases. It estimates the expected error rate in node t ($EE^{(t)}$) as follows:

$$EE^{(t)} = \min_{y_l} \left[\frac{N_x^{(t)} - N_{\bullet,y_l}^{(t)} + (1 - p_{\bullet,y_l}^{(t)}) \times m}{N_x^{(t)} + m} \right].$$ (2.39)

where m is a parameter that determines the importance of the a priori probability on the estimation of the error. Eq. (2.39), presented in [19], is a generalisation of the expected error rate presented in [82] if we assume that $m = k$ and that $p_{\bullet,y_l}^{(t)} = 1/k, \forall y_l \in Y$.

MEP is performed by comparing $EE^{(t)}$ with the weighted sum of the expected error rate of all children nodes from t. Each weight is given by $p_{v_j,\bullet}$, assuming v_j is the partition corresponding to the jth child of t. A disadvantage of MEP is the need of setting the ad-hoc parameter m. Usually, the higher the value of m, the more severe the pruning. Cestnik and Bratko [19] suggest that a domain expert should set m according to the level of noise in the data. Alternatively, a set of trees pruned with different values of m could be offered to the domain expert, so he/she can choose the best one according to his/her experience.

2.3.3.4 Critical-Value Pruning

The critical-value pruning, proposed by Mingers [73], is quite similar to the pre-pruning strategy *criterion 5* in Sect. 2.3.2. It is a bottom-up procedure that prunes a given non-terminal node t in case the value of its splitting measure is below a pre-determined threshold cv. Mingers [73] proposes the following two-step procedure for performing CVP:

1. Prune T for increasing values of cv, generating a set of pruned trees;
2. Choose the best tree among the set of trees (that includes T) by measuring each tree's accuracy (based on a pruning set) and significance (through the previously presented G statistic).

The disadvantage of CVP is the same of REP—the need of a pruning set. In addition, CVP does not present the optimality property that REP does, so there is no guarantee that the best tree found in step 2 is the smallest optimally pruned subtree of T, since the pruning step was performed based on the training set.

2.3.3.5 Cost-Complexity Pruning

Cost-complexity pruning is the post-pruning strategy of the CART system, detailed in [12]. It consists of two steps:

1. Generate a sequence of increasingly smaller trees, beginning with T and ending with the root node of T, by successively pruning the subtree yielding the lowest *cost complexity*, in a bottom-up fashion;
2. Choose the best tree among the sequence based on its relative size and accuracy (either on a pruning set, or provided by a cross-validation procedure in the training set).

The idea within step 1 is that pruned tree T_{i+1} is obtained by pruning the subtrees that show the lowest increase in the apparent error (error in the training set) per pruned leaf. Since the apparent error of pruned node t increases by the amount $r^{(t)} - r^{T^{(t)}}$, whereas its number of leaves decreases by $|\lambda_{T^{(t)}}| - 1$ units, the following ratio measures the increase in apparent error rate per pruned leaf:

$$\alpha = \frac{r^{(t)} - r^{T^{(t)}}}{|\lambda_{T^{(t)}}| - 1} \tag{2.40}$$

Therefore, T_{i+1} is obtained by pruning all nodes in T_i with the lowest value of α. T_0 is obtained by pruning all nodes in T whose α value is 0. It is possible to show that each tree T_i is associated to a distinct value α_i, such that $\alpha_i < \alpha_{i+1}$. Building the sequence of trees in step 1 takes quadratic time with respect to the number of internal nodes.

Regarding step 2, CCP chooses the smallest tree whose error (either on the pruning set or on cross-validation) is not more than one standard error (SE) greater than the

Fig. 2.2 Grafting of subtree
rooted in 4 onto the place of
parent 2. In **a** the original tree
T and in **b** the pruned tree T'

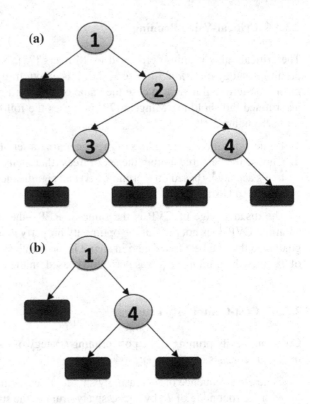

lowest error observed in the sequence of trees. This strategy is known as "1-SE"
variant since the work of Esposito et al. [33], which proposes ignoring the standard
error constraint, calling the strategy of selecting trees based only on accuracy of
"0-SE". It is argued that 1-SE has a tendency of overpruning trees, since its selection
is based on a conservative constraint [32, 33].

2.3.3.6 Error-Based Pruning

This strategy was proposed by Quinlan and it is implemented as the default pruning
strategy of C4.5 [89]. It is an improvement over PEP, based on a far more pessimistic
estimate of the expected error. Unlike PEP, EBP performs a bottom-up search, and
it performs not only the replacement of non-terminal nodes by leaves but also the
grafting[4] of subtree $T^{(t)}$ onto the place of parent t. Grafting is exemplified in Fig. 2.2.

Since grafting is potentially a time-consuming task, only the child subtree $T^{(t')}$ of
t with the greatest number of instances is considered to be grafted onto the place of t.

---

[4] *Grafting* is a term introduced by Esposito et al. [34]. It is also known as *subtree raising* [127].

For deciding whether to replace a non-terminal node by a leaf (subtree replacement), to graft a subtree onto the place of its parent (subtree raising) or not to prune at all, a pessimistic estimate of the expected error is calculated by using an upper confidence bound. Assuming that errors in the training set are binomially distributed with a given probability p in $N_x^{(t)}$ trials, it is possible to compute the exact value of the upper confidence bound as the value of p for which a binomially distributed random variable P shows $E^{(t)}$ successes in $N_x^{(t)}$ trials with probability CF. In other words, given a particular confidence CF (C4.5 default value is $CF = 25\%$), we can find the upper bound of the expected error (EE_{UB}) as follows:

$$EE_{UB} = \frac{f + \frac{z^2}{2N_x} + z\sqrt{\frac{f}{N_x} - \frac{f^2}{N_x} + \frac{z^2}{4N_x^2}}}{1 + \frac{z^2}{N_x}} \tag{2.41}$$

where $f = E^{(t)}/N_x$ and z is the number of standard deviations corresponding to the confidence CF (e.g., for $CF = 25\%$, $z = 0.69$).

In order to calculate the expected error of node t ($EE^{(t)}$), one must simply compute $N_x^{(t)} \times EE_{UB}$. For evaluating a subtree $T^{(t)}$, one must sum the expected error of every leaf of that subtree, i.e., $\sum_{s \in \lambda_{T^{(t)}}} EE^{(s)}$. Hence, given a non-terminal node t, it is possible to decide whether one should perform subtree replacement (when condition $EE^{(t)} \leq EE^{T^{(t)}}$ holds), subtree raising (when conditions $\exists j \in \zeta_t, EE^{(j)} < EE^{(t)} \wedge \forall i \in \zeta_t, N_x^{(i)} < N_x^{(j)}$ hold), or not to prune t otherwise.

An advantage of EBP is the new *grafting* operation that allows pruning useless branches without ignoring interesting lower branches (an elegant solution to the horizon effect problem). A drawback of the method is the parameter CF, even though it represents a confidence level. Smaller values of CF result in more pruning.

2.3.3.7 Empirical Evaluations

Some studies in the literature performed empirical analyses for evaluating pruning strategies. For instance, Quinlan [94] compared four methods of tree pruning (three of them presented in the previous sections—REP, PEP and CCP 1-SE). He argued that those methods in which a pruning set is needed (REP and CCP) did not perform noticeably better than the other methods, and thus their requirement for additional data is a weakness.

Mingers [71] compared five pruning methods, all of them presented in the previous sections (CCP, CVP, MEP, REP and PEP), and related them to different splitting measures. He states that pruning can improve the accuracy of induced decision trees by up to 25% in domains with noise and residual variation. In addition, he highlights the following findings: (i) MEP (the original version by Niblett and Bratko [82]) is the least accurate method due to its sensitivity to the number of classes in the data; (ii) PEP is the most "crude" strategy, though the fastest one—due to some bad results, it should be used with caution; (iii) CVP, CCP and REP performed well, providing consistently low error-rates for all data sets used; and (iv) there is no evidence of an interaction between the splitting measure and the pruning method used for inducing a decision tree.

Buntine [16], in his PhD thesis, also reports experiments on pruning methods (PEP, MEP, CCP 0-SE and 1-SE for both pruning set and cross-validation). Some of his findings were: (i) CCP 0-SE versions were marginally superior than the 1-SE versions; (ii) CCP 1-SE versions were superior in data sets with little apparent structure, where more severe pruning was inherently better; (iii) CCP 0-SE with cross-validation was marginally better than the other methods, though not in all data sets; and (iv) PEP performed reasonably well in all data sets, and was significantly superior in well-structured data sets (mushroom, glass and LED, all from UCI [36]);

Esposito et al. [32] compare the six post-pruning methods presented in the previous sections within an extended C4.5 system. Their findings were the following: (i) MEP, CVP, and EBP tend to underprune, whereas 1-SE (both cross-validation and pruning set versions) and REP have a propensity for overpruning; (ii) using a pruning-set is not usually a good option; (iii) PEP and EBP behave similarly, despite the difference in their formulation; (iv) pruning does not generally decrease the accuracy of a decision tree (only one of the domains tested was deemed as "pruning-averse"); and (v) data sets not prone to pruning are usually the ones with the highest base error whereas data sets with a low base error tend to benefit of any pruning strategy.

For a comprehensive survey of strategies for simplifying decision trees, please refer to [13]. For more details on post-pruning techniques in decision trees for regression, we recommend [12, 54, 85, 97, 113–115].

2.3.4 Missing Values

Handling missing values (denoted in this section by "?") is an important task not only in machine learning itself, but also in decision-tree induction. Missing values can be an issue during tree induction and also during classification. During tree induction, there are two moments in which we need to deal with missing values: splitting criterion evaluation and instances splitting.

During the split criterion evaluation in node t based on attribute a_i, some common strategies are:

- Ignore all instances belonging to the set $M = \{x_j | a_i(x_j) =?\}$ [12, 38];
- Imputation of missing values with the mode (nominal attributes) or the mean/median (numeric attributes) of all instances in t [24];
- Weight the splitting criterion value (calculated in node t with regard to a_i) by the proportion of missing values, i.e., $|M|/N_x^{(t)}$ [95].
- Imputation of missing values with the mode (nominal attributes) or the mean/median (numeric attributes) of all instances in t whose class attribute is the same of the instance whose a_i value is being imputed [65].

For deciding which child node training instance x_j should go to, considering a split in node t over a_i, and that $a_i(x_j) = ?$, some possibilities are:

- Ignore instance x_j [92];
- Treat instance x_j as if it has the most common value of a_i (mode or mean/median) [95];
- Weight instance x_j by the proportion of cases with known value in a given partition, i.e., $N_x^{(v_l)}/(N_x^{(t)} - |M|)$ (assuming t is the parent node and v_l is its lth partition) [57];
- Assign instance x_j to all partitions [38];
- Build an exclusive partition for missing values [95].
- Assign instance x_j to the partition with the greatest number of instances that belong to the same class that x_j. Formally, if x_j is labeled as y_l, we assign x_j to $\arg\max_{v_m}[N_{v_m,y_l}]$ [65].
- Create a *surrogate split* for each split in the original tree based on a different attribute [12]. For instance, a split over attribute a_i will have a surrogate split over attribute a_j, given that a_j is the attribute which most resembles the original split. Resemblance between two attributes in a binary tree is given by:

$$res(a_i, a_j, \mathbf{X}) = \frac{|\mathbf{X}_{a_i \in d_1(a_i) \wedge a_j \in d_1(a_j)}|}{N_x} + \frac{|\mathbf{X}_{a_i \in d_2(a_i) \wedge a_j \in d_2(a_j)}|}{N_x} \quad (2.42)$$

where the original split over attribute a_i is divided in two partitions, $d_1(a_i)$ and $d_2(a_i)$, and the alternative split over a_j is divided in $d_1(a_j)$ and $d_2(a_j)$. Hence, for creating a surrogate split, one must find attribute a_j that, after divided by two partitions $d_1(a_j)$ and $d_2(a_j)$, maximizes $res(a_i, a_j, \mathbf{X})$.

Finally, for classifying an unseen test instance x_j, considering a split in node t over a_i, and that $a_i(x_j) = ?$, some alternatives are:

- Explore all branches of t combining the results. More specifically, navigate through all ambiguous branches of the tree until reaching different leaves and choose class k with the highest probability, i.e., $\arg\max_y[\sum_{s \in \lambda_t}[N_{\bullet,y_l}^{(s)}]/N_x^{(t)}]$ [90];
- Treat instance x_j as if it has the most common value of a_i (mode or mean/median);
- Halt the classification process and assign instance x_j to the majority class of node t [95].

2.4 Other Induction Strategies

We presented a thorough review of the greedy top-down strategy for induction of decision trees in the previous section. In this section, we briefly present alternative strategies for inducing decision trees.

Bottom-up induction of decision trees was first mentioned in [59]. The authors propose a strategy that resembles agglomerative hierarchical clustering. The algorithm starts with each leaf having objects of the same class. In that way, a k-class problem will generate a decision tree with k leaves. The key idea is to merge, recursively, the two most similar classes in a non-terminal node. Then, a hyperplane is associated to the new non-terminal node, much in the same way as in top-down induction of oblique trees (in [59], a linear discriminant analysis procedure generates the hyperplanes). Next, all objects in the new non-terminal node are considered to be members of the same class (an artificial class that embodies the two *clustered* classes), and the procedure evaluates once again which are the two most similar classes. By recursively repeating this strategy, we end up with a decision tree in which the more obvious discriminations are done first, and the more subtle distinctions are postponed to lower levels. Landeweerd et al. [59] propose using the *Mahalanobis distance* to evaluate similarity among classes:

$$dist_M(i, j)^2 = (\mu_{\mathbf{y_i}} - \mu_{\mathbf{y_j}})^T \Sigma^{-1} (\mu_{\mathbf{y_i}} - \mu_{\mathbf{y_j}}) \qquad (2.43)$$

where $\mu_{\mathbf{y_i}}$ is the mean attribute vector of class y_i and Σ is the covariance matrix pooled over all classes.

Some obvious drawbacks of this strategy of bottom-up induction are: (i) binary-class problems provide a 1-level decision tree (root node and two children); such a simple tree cannot model complex problems; (ii) instances from the same class may be located in very distinct regions of the attribute space, harming the initial assumption that instances from the same class should be located in the same leaf node; (iii) hierarchical clustering and hyperplane generation are costly operations; in fact, a procedure for inverting the covariance matrix in the Mahalanobis distance is usually of time complexity proportional to $O(n^3)$.[5] We believe these issues are among the main reasons why bottom-up induction has not become as popular as top-down induction. For alleviating these problems, Barros et al. [4] propose a bottom-up induction algorithm named BUTIA that combines EM clustering with SVM classifiers. The authors later generalize BUTIA to a framework for generating oblique decision trees, namely BUTIF [5], which allows the application of different clustering and classification strategies.

Hybrid induction was investigated in [56]. The ideia is to combine both bottom-up and top-down approaches for building the final decision tree. The algorithm starts by executing the bottom-up approach as described above until two subgroups are achieved. Then, two centers (mean attribute vectors) and covariance information are extracted from these subgroups and used for dividing the training data in a top-down fashion according to a normalized sum-of-squared-error criterion. If the two new partitions induced account for separated classes, then the hybrid induction is finished; otherwise, for each subgroup that does not account for a class, recursively executes the hybrid induction by once again starting with the bottom-up procedure. Kim and Landgrebe [56] argue that in hybrid induction "It is more likely to converge to classes of informational value, because the clustering initialization provides early

---

[5] For inverting a matrix, the Gauss-Jordan procedure takes time proportional to $O(n^3)$. The fastest algorithm for inverting matrices to date is $O(n^{2.376})$ (the Coppersmith-Winograd algorithm).

guidance in that direction, while the straightforward top-down approach does not guarantee such convergence".

Several studies attempted on avoiding the greedy strategy usually employed for inducing trees. For instance, *lookahead* was employed for trying to improve greedy induction [17, 23, 29, 79, 84]. Murthy and Salzberg [79] show that one-level looka-head does not help building significantly better trees and can actually worsen the quality of trees induced. A more recent strategy for avoiding greedy decision-tree induction is to generate decision trees through *evolutionary algorithms*. The idea involved is to consider each decision tree as an individual in a population, which is evolved through a certain number of generations. Decision trees are modified by genetic operators, which are performed stochastically. A thorough review of decision-tree induction through evolutionary algorithms is presented in [6].

In a recent work, Basgalupp et al. [7] propose a decision-tree induction algorithm (called Beam Classifier) that seeks to avoid being trapped in local-optima by doing a beam search during the decision tree growth. A beam search algorithm keeps track of w states rather than just one. It begins with w randomly generated states (decision trees). At each step, all the successors of the w states are generated. If any successor is a goal state, the algorithm halts. Otherwise, it selects the w best successors from the complete list, discards the other states in the list, and repeats this loop until the quality of the best current tree cannot be improved. An interesting fact regarding the beam search algorithm is that if we set $w = 1$ we are actually employing the greedy strategy for inducing decision trees.

Beam Classifier starts with n empty decision trees (root nodes), where n is the number of data set attributes, and each root node represents an attribute. Then, the algorithm selects the best w trees according to a given criterion, and each one is expanded. For each expansion, the algorithm performs an adapted pre-order tree search method, expanding recursively (for each attribute) the leaf nodes from left to right. Thus, this expansion results in t new trees,

$$t = \sum_{i=1}^{w} \sum_{j=1}^{|\lambda_i|} m_{ij} \qquad (2.44)$$

where w is the beam-width and m_{ij} is the number of available attributes[6] at the jth leaf node of the ith tree. Then, the algorithm selects once again the best w trees considering a pool of p trees, $p = w + t$. This process is repeated until a stop criterion is satisfied.

Other examples of non-greedy strategies for inducing decision trees include: (i) using *linear programming* to complement greedy-induced decision trees [8]; (ii) incremental and non-incremental *restructuring* of decision trees [118]; (iii) *skewing* the data to simulate an alternative distribution in order to deal with problematic cases for decision trees (e.g., the parity-like function) [86]; and (iv) *anytime learning* of decision trees [31].

---

[6] Nominal attributes are not used more than once in a given subtree.

Even though a lot of effort has been employed in the design of a non-greedy decision-tree induction algorithm, it is still debatable whether the proposed attempts can consistently obtain better results than the greedy top-down framework. Most of the times, the gain in performance obtained with a non-greedy approach is not sufficient to compensate for the extra computational effort.

2.5 Chapter Remarks

In this chapter, we presented the main design choices one has to face when programming a decision-tree induction algorithm. We gave special emphasis to the greedy top-down induction strategy, since it is by far the most researched technique for decision-tree induction.

Regarding top-down induction, we presented the most well-known splitting measures for univariate decision trees, as well as some new criteria found in the literature, in an unified notation. Furthermore, we introduced some strategies for building decision trees with multivariate tests, the so-called *oblique* trees. In particular, we showed that efficient oblique decision-tree induction has to make use of heuristics in order to derive "good" hyperplanes within non-terminal nodes. We detailed the strategy employed in the OC1 algorithm [77, 80] for deriving hyperplanes with the help of a *randomized* perturbation process. Following, we depicted the most common stopping criteria and post-pruning techniques employed in classic algorithms such as CART [12] and C4.5 [89], and we ended the discussion on top-down induction with an enumeration of possible strategies for dealing with missing values, either in the growing phase or during classification of a new instance.

We ended our analysis on decision trees with some alternative induction strategies, such as bottom-up induction and hybrid-induction. In addition, we briefly discussed work that attempt to avoid the greedy strategy, by either implementing *lookahead* techniques, *evolutionary algorithms*, *beam-search*, *linear programming*, *(non-) incremental restructuring*, *skewing*, or *anytime learning*. In the next chapters, we present an overview of *evolutionary algorithms* and *hyper-heuristics*, and review how they can be applied to decision-tree induction.

References

1. A. Agresti, *Categorical Data Analysis*, 2nd edn., Wiley Series in Probability and Statistics (Wiley-Interscience, Hoboken, 2002)
2. E. Alpaydin, *Introduction to Machine Learning* (MIT Press, Cambridge, 2010). ISBN: 026201243X, 9780262012430
3. P.W. Baim, A method for attribute selection in inductive learning systems. IEEE Trans. Pattern Anal. Mach. Intell. **10**(6), 888–896 (1988)
4. R.C. Barros et al., A bottom-up oblique decision tree induction algorithm, in *11th International Conference on Intelligent Systems Design and Applications*, pp. 450–456 (2011)

5. R.C. Barros et al., A framework for bottom-up induction of decision trees, Neurocomputing (2013 in press)
6. R.C. Barros et al., A survey of evolutionary algorithms for decision-tree induction. IEEE Trans. Syst. Man, Cybern. Part C: Appl. Rev. **42**(3), 291–312 (2012)
7. M.P. Basgalupp et al., A beam-search based decision-tree induction algorithm, in *Machine Learning Algorithms for Problem Solving in Computational Applications: Intelligent Techniques*. IGI-Global (2011)
8. K. Bennett, Global tree optimization: a non-greedy decision tree algorithm. Comput. Sci. Stat. **26**, 156–160 (1994)
9. K. Bennett, O. Mangasarian, Multicategory discrimination via linear programming. Optim. Methods Softw. **2**, 29–39 (1994)
10. K. Bennett, O. Mangasarian, Robust linear programming discrimination of two linearly inseparable sets. Optim. Methods Softw. **1**, 23–34 (1992)
11. L. Bobrowski, M. Kretowski, Induction of multivariate decision trees by using dipolar criteria, in *European Conference on Principles of Data Mining and Knowledge Discovery*. pp. 331–336 (2000)
12. L. Breiman et al., *Classification and Regression Trees* (Wadsworth, Belmont, 1984)
13. L. Breslow, D. Aha, Simplifying decision trees: a survey. Knowl. Eng. Rev. **12**(01), 1–40 (1997)
14. C.E. Brodley, P.E. Utgoff, *Multivariate versus univariate decision trees*. Technical Report. Department of Computer Science, University of Massachusetts at Amherst (1992)
15. A. Buja, Y.-S. Lee, Data mining criteria for tree-based regression and classification, in *ACM SIGKDD International Conference on Knowledge Discovery and Data Mining*. pp. 27–36 (2001)
16. W. Buntine, A theory of learning classification rules, PhD thesis. University of Technology, Sydney (1992)
17. W. Buntine, Learning classification trees. Stat. Comput. **2**, 63–73 (1992)
18. R. Casey, G. Nagy, Decision tree design using a probabilistic model. IEEE Trans. Inf. Theory **30**(1), 93–99 (1984)
19. B. Cestnik, I. Bratko, On estimating probabilities in tree pruning, *Machine Learning-EWSL-91*, Vol. 482. Lecture Notes in Computer Science (Springer, Berlin, 1991), pp. 138–150
20. B. Chandra, R. Kothari, P. Paul, A new node splitting measure for decision tree construction. Pattern Recognit. **43**(8), 2725–2731 (2010)
21. B. Chandra, P.P. Varghese, Moving towards efficient decision tree construction. Inf. Sci. **179**(8), 1059–1069 (2009)
22. J. Ching, A. Wong, K. Chan, Class-dependent discretization for inductive learning from continuous and mixed-mode data. IEEE Trans. Pattern Anal. Mach. Intell. **17**(7), 641–651 (1995)
23. P. Chou, Optimal partitioning for classification and regression trees. IEEE Trans. Pattern Anal. Mach. Intell. **13**(4), 340–354 (1991)
24. P. Clark, T. Niblett, The CN2 induction algorithm. Mach. Learn. **3**(4), 261–283 (1989)
25. D. Coppersmith, S.J. Hong, J.R.M. Hosking, Partitioning nominal attributes in decision trees. Data Min. Knowl. Discov. **3**, 197–217 (1999)
26. R.L. De Mántaras, A distance-based attribute selection measure for decision tree induction. Mach. Learn. **6**(1), 81–92 (1991). ISSN: 0885–6125
27. G. De'ath, Multivariate regression trees: A new technique for modeling species-environment relationships. Ecology **83**(4), 1105–1117 (2002)
28. L. Devroye, L. Györfi, G. Lugosi, *A Probabilistic Theory of Pattern Recognition* (Springer, New York, 1996)
29. M. Dong, R. Kothari, Look-ahead based fuzzy decision tree induction. IEEE Trans. Fuzzy Syst. **9**(3), 461–468 (2001)
30. B. Draper, C. Brodley, Goal-directed classification using linear machine decision trees. IEEE Trans. Pattern Anal. Mach. Intell. **16**(9), 888–893 (1994)

31. S. Esmeir, S. Markovitch, Anytime learning of decision trees. J. Mach. Learn. Res. **8**, 891–933 (2007)
32. F. Esposito, D. Malerba, G. Semeraro, A comparative analysis of methods for pruning decision trees. IEEE Trans. Pattern Anal. Mach. Intell. **19**(5), 476–491 (1997)
33. F. Esposito, D. Malerba, G. Semeraro, A further study of pruning methods in decision tree induction, in *Fifth International Workshop on Artificial Intelligence and Statistics*. pp. 211–218 (1995)
34. F. Esposito, D. Malerba, G. Semeraro, Simplifying decision trees by pruning and grafting: new results (extended abstract), in *8th European Conference on Machine Learning*. ECML'95. (Springer, London, 1995) pp. 287–290
35. U. Fayyad, K. Irani, The attribute selection problem in decision tree generation, in *National Conference on Artificial Intelligence*. pp. 104–110 (1992)
36. A. Frank, A. Asuncion, *UCI Machine Learning Repository* (2010)
37. A.A. Freitas, A critical review of multi-objective optimization in data mining: a position paper. SIGKDD Explor. Newsl. **6**(2), 77–86 (2004). ISSN: 1931–0145
38. J.H. Friedman, A recursive partitioning decision rule for nonparametric classification. IEEE Trans. Comput. **100**(4), 404–408 (1977)
39. S.B. Gelfand, C.S. Ravishankar, E.J. Delp, An iterative growing and pruning algorithm for classification tree design. IEEE Int. Conf. Syst. Man Cybern. **2**, 818–823 (1989)
40. M.W. Gillo, MAID: A Honeywell 600 program for an automatised survey analysis. Behav. Sci. **17**, 251–252 (1972)
41. M. Gleser, M. Collen, Towards automated medical decisions. Comput. Biomed. Res. **5**(2), 180–189 (1972)
42. L.A. Goodman, W.H. Kruskal, Measures of association for cross classifications. J. Am. Stat. Assoc. **49**(268), 732–764 (1954)
43. T. Hancock et al., Lower bounds on learning decision lists and trees. Inf. Comput. **126**(2) (1996)
44. C. Hartmann et al., Application of information theory to the construction of efficient decision trees. IEEE Trans. Inf. Theory **28**(4), 565–577 (1982)
45. R. Haskell, A. Noui-Mehidi, Design of hierarchical classifiers, in *Computing in the 90s*, Vol. 507. Lecture Notes in Computer Science, ed. by N. Sherwani, E. de Doncker, J. Kapenga (Springer, Berlin, 1991), pp. 118–124
46. H. Hauska, P. Swain, The decision tree classifier: design and potential, in *2nd Symposium on Machine Processing of Remotely Sensed Data* (1975)
47. D. Heath, S. Kasif, S. Salzberg, Induction of oblique decision trees. J. Artif. Intell. Res. **2**, 1–32 (1993)
48. W. Hsiao, Y. Shih, Splitting variable selection for multivariate regression trees. Stat. Probab. Lett. **77**(3), 265–271 (2007)
49. E.B. Hunt, J. Marin, P.J. Stone, *Experiments in Induction* (Academic Press, New York, 1966)
50. L. Hyafil, R. Rivest, Constructing optimal binary decision trees is NP-complete. Inf. Process. Lett. **5**(1), 15–17 (1976)
51. A. Ittner, Non-linear decision trees, in *13th International Conference on Machine Learning*. pp. 1–6 (1996)
52. B. Jun et al., A new criterion in selection and discretization of attributes for the generation of decision trees. IEEE Trans. Pattern Anal. Mach. Intell. **19**(2), 1371–1375 (1997)
53. G. Kalkanis, The application of confidence interval error analysis to the design of decision tree classifiers. Pattern Recognit. Lett. **14**(5), 355–361 (1993)
54. A. Karalič, Employing linear regression in regression tree leaves, *10th European Conference on Artificial Intelligence*. ECAI'92 (Wiley, New York, 1992)
55. G.V. Kass, An exploratory technique for investigating large quantities of categorical data. APPL STATIST **29**(2), 119–127 (1980)
56. B. Kim, D. Landgrebe, Hierarchical classifier design in high-dimensional numerous class cases. IEEE Trans. Geosci. Remote Sens. **29**(4), 518–528 (1991)

57. I. Kononenko, I. Bratko, E. Roskar, Experiments in automatic learning of medical diagnostic rules. Technical Report Ljubljana, Yugoslavia: Jozef Stefan Institute (1984)
58. I. Kononenko, Estimating attributes: analysis and extensions of RELIEF, *Proceedings of the European Conference on Machine Learning on Machine Learning* (Springer, New York, 1994). ISBN: 3-540-57868-4
59. G. Landeweerd et al., Binary tree versus single level tree classification of white blood cells. Pattern Recognit. **16**(6), 571–577 (1983)
60. D.R. Larsen, P.L. Speckman, Multivariate regression trees for analysis of abundance data. Biometrics **60**(2), 543–549 (2004)
61. Y.-S. Lee, A new splitting approach for regression trees. Technical Report. Dongguk University, Department of Statistics: Dongguk University, Department of Statistics (2001)
62. X. Li, R.C. Dubes, Tree classifier design with a permutation statistic. Pattern Recognit. **19**(3), 229–235 (1986)
63. X.-B. Li et al., Multivariate decision trees using linear discriminants and tabu search. IEEE Trans. Syst., Man, Cybern.-Part A: Syst. Hum. **33**(2), 194–205 (2003)
64. H. Liu, R. Setiono, Feature transformation and multivariate decision tree induction. Discov. Sci. **1532**, 279–291 (1998)
65. W. Loh, Y. Shih, Split selection methods for classification trees. Stat. Sin. **7**, 815–840 (1997)
66. W. Loh, Regression trees with unbiased variable selection and interaction detection. Stat. Sin. **12**, 361–386 (2002)
67. D. Malerba et al., Top-down induction of model trees with regression and splitting nodes. IEEE Trans. Pattern Anal. Mach. Intell. **26**(5), 612–625 (2004)
68. O. Mangasarian, R. Setiono, W. H. Wolberg, Pattern recognition via linear programming: theory and application to medical diagnosis, in *SIAM Workshop on Optimization* (1990)
69. N. Manwani, P. Sastry, A Geometric Algorithm for Learning Oblique Decision Trees, in *Pattern Recognition and Machine Intelligence*, ed. by S. Chaudhury, et al. (Springer, Berlin, 2009), pp. 25–31
70. J. Martin, An exact probability metric for decision tree splitting and stopping. Mach. Learn. **28**(2), 257–291 (1997)
71. J. Mingers, An empirical comparison of pruning methods for decision tree induction. Mach. Learn. **4**(2), 227–243 (1989)
72. J. Mingers, An empirical comparison of selection measures for decision-tree induction. Mach. Learn. **3**(4), 319–342 (1989)
73. J. Mingers, Expert systems—rule induction with statistical data. J. Oper. Res. Soc. **38**, 39–47 (1987)
74. T.M. Mitchell, *Machine Learning* (McGraw-Hill, New York, 1997)
75. F. Mola, R. Siciliano, A fast splitting procedure for classification trees. Stat. Comput. **7**(3), 209–216 (1997)
76. J.N. Morgan, R.C. Messenger, *THAID: a sequential search program for the analysis of nominal scale dependent variables*. Technical Report. Institute for Social Research, University of Michigan (1973)
77. S.K. Murthy, S. Kasif, S.S. Salzberg, A system for induction of oblique decision trees. J. Artif. Intell. Res. **2**, 1–32 (1994)
78. S.K. Murthy, Automatic construction of decision trees from data: A multi-disciplinary survey. Data Min. Knowl. Discov. **2**(4), 345–389 (1998)
79. S.K. Murthy, S. Salzberg, Lookahead and pathology in decision tree induction, in *14th International Joint Conference on Artificial Intelligence*. (Morgan Kaufmann, San Francisco, 1995), pp. 1025–1031
80. S.K. Murthy et al., OC1: A randomized induction of oblique decision trees, in *Proceedings of the 11th National Conference on Artificial Intelligence* (AAAI'93), pp. 322–327 (1993)
81. G.E. Naumov, NP-completeness of problems of construction of optimal decision trees. Sov. Phys. Doklady **36**(4), 270–271 (1991)
82. T. Niblett, I. Bratko, Learning decision rules in noisy domains, in *6th Annual Technical Conference on Research and Development in Expert Systems III*. pp. 25–34 (1986)

83. N.J. Nilsson, *The Mathematical Foundations of Learning Machines* (Morgan Kaufmann Publishers Inc., San Francisco, 1990). ISBN: 1-55860-123-6
84. S.W. Norton, Generating better decision trees, *11th International Joint Conference on Artificial Intelligence* (Morgan Kaufmann Publishers Inc., San Francisco, 1989)
85. K. Osei-Bryson, Post-pruning in regression tree induction: an integrated approach. Expert Syst. Appl. **34**(2), 1481–1490 (2008)
86. D. Page, S. Ray, Skewing: An efficient alternative to lookahead for decision tree induction, in *18th International Joint Conference on Artificial Intelligence* (Morgan Kaufmann Publishers Inc., San Francisco, 2003), pp. 601–607
87. A. Patterson, T. Niblett, *ACLS User Manual* (Intelligent Terminals Ltd., Glasgow, 1983)
88. K. Pattipati, M. Alexandridis, Application of heuristic search and information theory to sequential fault diagnosis. IEEE Trans. Syst. Man Cybern. **20**, 872–887 (1990)
89. J.R. Quinlan, *C4.5: Programs for Machine Learning* (Morgan Kaufmann, San Francisco, 1993). ISBN: 1-55860-238-0
90. J.R. Quinlan, Decision trees as probabilistic classifiers, in *4th International Workshop on Machine Learning* (1987)
91. J.R. Quinlan, Discovering rules by induction from large collections of examples, in *Expert Systems in the Micro-elect Age*, ed. by D. Michie (Edinburgh University Press, Edinburgh, 1979)
92. J.R. Quinlan, Induction of decision trees. Mach. Learn. **1**(1), 81–106 (1986)
93. J.R. Quinlan, Learning with continuous classes, in *5th Australian Joint Conference on Artificial Intelligent.* **92**, pp. 343–348 (1992)
94. J.R. Quinlan, Simplifying decision trees. Int. J. Man-Mach. Stud. **27**, 221–234 (1987)
95. J.R. Quinlan, Unknown attribute values in induction, in *6th International Workshop on Machine Learning.* pp. 164–168 (1989)
96. J.R. Quinlan, R.L. Rivest, Inferring decision trees using the minimum description length principle. Inf. Comput. **80**(3), 227–248 (1989)
97. M. Robnik-Sikonja, I. Kononenko, Pruning regression trees with MDL, in *European Conference on Artificial Intelligence.* pp. 455–459 (1998)
98. L. Rokach, O. Maimon, Top-down induction of decision trees classifiers—a survey. IEEE Trans. Syst. Man, Cybern. Part C: Appl. Rev. **35**(4), 476–487 (2005)
99. E.M. Rounds, A combined nonparametric approach to feature selection and binary decision tree design. Pattern Recognit. **12**(5), 313–317 (1980)
100. J.P. Sá et al., Decision trees using the minimum entropy-of-error principle, in *13th International Conference on Computer Analysis of Images and Patterns* (Springer, Berlin, 2009), pp. 799–807
101. S. Safavian, D. Landgrebe, A survey of decision tree classifier methodology. IEEE Trans. Syst. Man Cybern. **21**(3), 660–674 (1991). ISSN: 0018–9472
102. I.K. Sethi, G.P.R. Sarvarayudu, Hierarchical classifier design using mutual information. IEEE Trans. Pattern Anal. Mach. Intell. **4**(4), 441–445 (1982)
103. S. Shah, P. Sastry, New algorithms for learning and pruning oblique decision trees. IEEE Trans. Syst. Man, Cybern. Part C: Applic. Rev. **29**(4), 494–505 (1999)
104. C.E. Shannon, A mathematical theory of communication. BELL Syst. Tech. J. **27**(1), 379–423, 625–56 (1948)
105. Y. Shih, Selecting the best categorical split for classification trees. Stat. Probab. Lett. **54**, 341–345 (2001)
106. L.M. Silva et al., Error entropy in classification problems: a univariate data analysis. Neural Comput. **18**(9), 2036–2061 (2006)
107. J.A. Sonquist, E.L. Baker, J.N. Morgan, *Searching for structure*. Technical Report. Institute for Social Research University of Michigan (1971)
108. J. Talmon, A multiclass nonparametric partitioning algorithm. Pattern Recognit. Lett. **4**(1), 31–38 (1986)
109. P.J. Tan, D.L. Dowe, MML inference of oblique decision trees, in *17th Australian Joint Conference on AI.* pp. 1082–1088 (2004)

110. P.-N. Tan, M. Steinbach, V. Kumar, *Introduction to Data Mining* (Addison-Wesley, Boston, 2005)
111. P.C. Taylor, B.W. Silverman, Block diagrams and splitting criteria for classification trees. Stat. Comput. **3**, 147–161 (1993)
112. P. Taylor, M. Jones, Splitting criteria for regression trees. J. Stat. Comput. Simul. **55**(4), 267–285 (1996)
113. L. Torgo, Functional models for regression tree leaves, in *14th International Conference on Machine Learning*. ICML'97. (Morgan Kaufmann Publishers Inc., San Francisco, 1997), pp. 385–393
114. L. Torgo, A comparative study of reliable error estimators for pruning regression trees, in *Iberoamerican Conference on Artificial Intelligence* (Springer, Berlin, 1998), pp. 1–12
115. L. Torgo, Error estimators for pruning regression trees, in *10th European Conference on Machine Learning* (Springer, Berlin, 1998), pp. 125–130
116. K.P. Unnikrishnan, K.P. Venugopal, Alopex: A correlation-based learning algorithm for feed-forward and recurrent neural networks. Neural Comput. **6**, 469–490 (1994)
117. P.E. Utgoff, Perceptron trees: a case study in hybrid concept representations. Connect. Sci. **1**(4), 377–391 (1989)
118. P.E. Utgoff, N.C. Berkman, J.A. Clouse, Decision tree induction based on efficient tree restructuring. Mach. Learn. **29**(1), 5–44 (1997)
119. P.E. Utgoff, C.E. Brodley, *Linear machine decision trees*. Technical Report. University of Massachusetts, Dept of Comp Sci (1991)
120. P.E. Utgoff, J.A. Clouse. *A Kolmogorov-Smirnoff Metric for Decision Tree Induction*. Technical Report. University of Massachusetts, pp. 96–3 (1996)
121. P. Utgoff, C. Brodley, An incremental method for finding multivariate splits for decision trees, in *7th International Conference on Machine Learning*. pp. 58–65 (1990)
122. P.K. Varshney, C.R.P. Hartmann, J.M.J. de Faria, Application of information theory to sequential fault diagnosis. IEEE Trans. Comput. **31**(2), 164–170 (1982)
123. D. Wang, L. Jiang, An improved attribute selection measure for decision tree induction, *in: 4th International Conference on Fuzzy Systems and Knowledge Discovery*. pp. 654–658 (2007)
124. Y. Wang, I.H. Witten, Induction of model trees for predicting continuous classes, in *Poster papers of the 9th European Conference on Machine Learning* (Springer, Berlin, 1997)
125. A.P. White, W.Z. Liu, Technical note: Bias in information-based measures in decision tree induction. Mach. Learn. **15**(3), 321–329 (1994)
126. S.S. Wilks, *Mathematical Statistics* (Wiley, New York, 1962)
127. I.H. Witten, E. Frank, *Data Mining: Practical Machine Learning Tools and Techniques with Java Implementations* (Morgan Kaufmann, San Francisco, 1999). ISBN: 1558605525
128. C.T. Yildiz, E. Alpaydin, Omnivariate decision trees. IEEE Trans. Neural Netw. **12**(6), 1539–1546 (2001)
129. H. Zantema, H. Bodlaender, Finding small equivalent decision trees is hard. Int. J. Found. Comput. Sci. **11**(2), 343–354 (2000)
130. X. Zhou, T. Dillon, A statistical-heuristic feature selection criterion for decision tree induction. IEEE Trans. Pattern Anal. Mac. Intell. **13**(8), 834–841 (1991)

Chapter 3
Evolutionary Algorithms and Hyper-Heuristics

Abstract This chapter presents the basic concepts of evolutionary algorithms (EAs) and hyper-heuristics (HHs), which are computational techniques directly explored in this book. EAs are well-known population-based metaheuristics. They have been employed in artificial intelligence over several years with the goal of providing the near-optimal solution for a problem that comprises a very large search space. A general overview of EAs is presented in Sect. 3.1. HHs, in turn, are a recently new field in the optimisation research area, in which a metaheuristic—often an EA, and this is why these related concepts are reviewed together in this chapter—is used for searching in the space of heuristics (algorithms), and not in the space of solutions, like conventional metaheuristics. The near-optimal heuristic (algorithm) provided by a HHs approach can be further employed in several distinct problems, instead of relying on a new search process for each new problem to be solved. An overview of HHs is given in Sect. 3.2.

Keywords Evolutionary algorithms · Evolutionary computation · Metaheuristics · Hyper-heuristics

3.1 Evolutionary Algorithms

Evolutionary algorithms (EAs) are a collection of optimisation techniques whose design is based on metaphors of biological processes. Fretias [20] defines EAs as "stochastic search algorithms inspired by the process of neo-Darwinian evolution", and Weise [44] states that "EAs are population-based metaheuristic optimisation algorithms that use biology-inspired mechanisms (…) in order to refine a set of solution candidates iteratively".

The idea surrounding EAs is the following. There is a population of individuals, where each individual is a possible solution to a given problem. This population evolves towards increasingly better solutions through stochastic operators. After the evolution is completed, the fittest individual represents a "near-optimal" solution for the problem at hand.

© The Author(s) 2015
R.C. Barros et al., *Automatic Design of Decision-Tree Induction Algorithms*,
SpringerBriefs in Computer Science, DOI 10.1007/978-3-319-14231-9_3

For evolving individuals, an EA evaluates each individual through a fitness function that measures the quality of the solutions that are being evolved. After the evaluation of all individuals that are part of the initial population, the algorithm's iterative process starts. At each iteration, hereby called *generation*, the fittest individuals have a higher probability of being selected for reproduction to increase the chances of producing good solutions. The selected individuals undergo stochastic genetic operators, such as crossover and mutation, producing new offspring. These new individuals will replace the current population of individuals and the evolutionary process continues until a stopping criterion is satisfied (e.g., until a fixed number of generations is achieved, or until a satisfactory solution has been found).

There are several kinds of EAs, such as genetic algorithms (GAs), genetic programming (GP), classifier systems (CS), evolution strategies (ES), evolutionary programming (EP), estimation of distribution algorithms (EDA), etc. This chapter will focus on GA and GP, the most commonly used EAs for data mining [19]. At a high level of abstraction, GAs and GP can be described by the pseudocode in Algorithm 1.

Algorithm 1 Pseudo-code for GP and GAs.

1: Create initial population
2: Calculate fitness of each individual
3: **repeat**
4:       Select individuals based on fitness
5:       Apply genetic operators to selected individuals, creating new individuals
6:       Compute fitness of each new individual
7:       Update the current population
8: **until** (stopping criteria) **return** Best individual

GAs, which were initially presented by Holland in his pioneering monograph [25], are defined as:

> (...) search algorithms based on the mechanics of natural selection and natural genetics. They combine survival of the fittest among *string structures* [our italics] with a structured yet randomized information exchange to form a search algorithm with some of the innovative flair of human search (p. 1).

Representation is a key issue in GAs, and while they are capable of solving a great many problems, the use of fixed-length character strings may not work on a variety of cases. John Koza, the researcher responsible for spreading the GP concepts to the research community, argues in his text book on GP [26] that the initial selection of string length limits the number of internal states of the system and also limits what the system can learn. Moreover, he states that representation schemes based on fixed-length character strings do not provide a convenient way of representing computational procedures or of incorporating iteration or recursion when these capabilities are desirable or necessary to solve a given problem. Hence, he defines GP as:

(...) a paradigm that deals with the problem of representation in genetic algorithms by increasing the complexity of the structures undergoing adaptation. In particular, the structures undergoing adaptation in genetic programming are general, hierarchical computer programs of dynamically varying size and shape (p. 73).

After analysing the definitions of GAs and GP, it is not hard to see why the type of solution encoding in an EA is usually argued to determine the type of EA used. If solutions are encoded in a fixed-length linear string, researchers claim a GA is being used. Conversely, tree-encoding schemes usually imply the use of GP. Although solution encoding can differentiate between GAs and GP, the main question perhaps is not what the representation is (e.g. a linear string or a tree) but rather how the representation is interpreted [4].

In this sense, Woodward [45] recommends defining GAs and GP according to the genotype-phenotype mapping: if there is a one-to-one mapping, the EA in question is a GA; if there is a many-to-one mapping, the EA is a GP. Nevertheless, this definition is tricky. For instance, assume a feature selection problem in data mining, where an individual (chromosome) consists of n genes, one for each attribute. Now assume that each gene contains a real value in the range [0, 1], representing the probability of the corresponding attribute being selected. Assume also that, for decoding a chromosome, a threshold is predefined, and an attribute is selected only if the value of its gene is larger than that threshold. In this case, we have a many-to-one mapping, because there are many different genotypes (different arrays of probabilities) that may be decoded into the same phenotype (the same set of selected features). This particular many-to-one mapping does not indicate we are dealing with GP. Actually, we can use the same set of genetic operators and remaining parameters of a typical GA for this scenario. One may say that a good distinction between GAs and GP is whether a solution encodes data only (GAs) or data and functions (GP).

We present in Chap. 4 an EA that automatically designs decision-tree algorithms. It encodes its individuals as integer vectors and employs the same set of genetic operators and remaining parameters of a typical GA. However, it encodes both data (values of parameters) and functions (design components of decision-tree algorithms such as pruning methods). Given the difficulty in clearly defining whether the proposed approach is a GA or a GP algorithm, we henceforth adopt the generic term "evolutionary algorithm" when referring to our method. In the next few sections, we review the basic concepts involved when designing an EA: (i) individual representation and population initialization; (ii) fitness function; and (iii) selection methods and genetic operators.

3.1.1 Individual Representation and Population Initialization

The first step for designing an EA is to define how its individuals are going to be represented (encoded). Since each individual is a target solution to the problem at hand, so if we want to discover the best decision-tree induction algorithm for a given set of data sets, each individual should be a candidate decision-tree induction algorithm.

It is during the step of deciding the individual encoding scheme that a link must be created between the "real world" and the "evolutionary world" [16]. Thus, the encoding scheme creates the bridge between the original optimisation problem and the search space to be explored by the EA.

The objects that represent possible solutions within the context of the original problem are referred to as phenotypes, whereas their encoding (the EA's individuals) are referred to as genotypes. Note that the phenotype search space may differ significantly from the genotype search space. A genotype must be decoded so we can have a clear notion of its performance as a candidate solution to the original problem.

Defining a suitable individual encoding scheme is not an easy task, and it may be decisive for the success of the EA. As we will see in the next few sections, there is a direct relation between individual representation and the genetic operators that guide the evolutionary process. Finding suitable encoding schemes for the problem at hand is a task that usually comes with practical experience and solid knowledge on the application domain [31].

GA individuals, as previously seen, are encoded as fixed-length character strings such as binary or integer vectors. GP individuals, on the other hand, are usually represented in a tree-based scheme (a consequence of the GP definition by Koza). Notwithstanding, there are also studies that encode GP individuals as linear structures, and even graphs [3]. In Fig. 3.1, we have a GP individual that comprises four terminals (a, b, c, d) and three functions (OR, −, ×).

The initialization of individuals in a GA is usually completely random, i.e., a random value within the range of accepted values is generated for each gene (position of the vector) of the individual's genome. Nevertheless, domain knowledge may be inserted in order to guarantee some robustness to the individuals of the initial population.

The initialization of individuals in GP offers more possibilities due to its usually-adopted tree structure. Koza [26] proposes two distinct methods (both in the context of tree representation): full and *grow*. The full method works as follows: until the maximum initial depth of the tree is reached, randomly choose non-terminals nodes; when the maximum depth is reached, randomly choose terminal nodes. This method leads to trees whose branches are of equal size. Alternatively, the grow method allows the choice of either non-terminals and terminals (except for the root node which is always a non-terminal). Hence, the trees generated may have irregular shapes, because terminal nodes end the growth of the branch even if the maximum depth has not yet been reached. The grow method is said to have serious weakness, such as [27]: (i) the method picks all functions with equal likelihood; there is no way to fine tune the preference of certain functions over others; (ii) the method does not give the user much control over the tree structures generated; and (iii) while the maximum depth parameter (or, alternatively, the maximum number of nodes) is used as an upper bound on maximal tree depth, there is no appropriate way to create trees

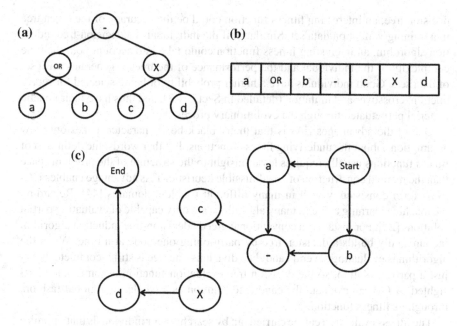

Fig. 3.1 A GP individual represented as: **a** a tree; **b** a linear structure and **c** a graph

with either a fixed or average tree size or depth. The full method, in turn, is criticized for producing a very narrow range of tree structures.

A third option to generate trees is the ramped half-and-half method (RHH). It is intended to enhance the diversity of individuals by generating, for each depth level, half of the trees through the grow method and half of the trees through the full method. Burke et al. [7] draw some considerations on the bias of the RHH method, though most of their analysis is inconclusive. The only known fact is that RHH tends to generate more trees through the full method. This occurs because duplicates are typically not allowed, and the grow method tends to generate smaller trees (hence it is more susceptible to generate duplicates).

Other tree-generation algorithms for genetic programming include *probabilistic tree-creation* 1 and 2 (PTC 1 and PTC 2) [27], *RandomBranch* [12] and *exact uniform initialization* [6], just to name a few.

3.1.2 Fitness Function

After initializing the population, an EA evaluates the goodness of each individual. A *fitness function* is responsible for evaluating how well an individual solves the target problem. It is mainly problem-dependent, i.e., different problems tend to demand different fitness functions. For instance, in a problem where each individual is a decision tree, an interesting fitness function could be the accuracy of the given tree in a training/validation data set. Similarly, if the individual is a decision-tree induction algorithm, an interesting fitness function could take into account both the time complexity of the individual and the performance of the trees it generates for a set of data sets. Good individuals have a higher probability of being selected for reproduction, crossover and mutation (detailed in Sect. 3.1.3), and thus have their genetic material perpetuated through the evolutionary process.

One of the advantages EAs is that their "black-box" character makes only few assumptions about the underlying fitness functions. In other words, the definition of fitness functions usually requires lesser insight to the structure of the problem space than the manual construction of an admissible heuristic. This advantage enables EAs to perform consistently well in many different problem domains [44]. Regarding the machine learning context, many algorithms are only capable of evaluating partial solutions [20]. For instance, a conventional greedy decision-tree induction algorithm incrementally builds a decision tree by partitioning one node at a time. When the algorithm is evaluating several candidate divisions, the tree is still incomplete, being just a partial solution, so the decision tree evaluation function is somewhat short-sighted. A GP can evaluate the candidate solution as a whole, in a global fashion, through its fitness function.

The fitness evaluation can be carried out by searching for individuals that optimise a single measure (i.e., single-objective evaluation), or by optimising multiple objectives (i.e., multi-objective evaluation). Multi-objective evaluation is quite useful in the context of machine learning. For instance, it is often desirable to obtain decision trees that are both accurate and comprehensible, in the shortest space of time. Several approaches for handling multi-objective problems were proposed in the literature, such as the weighted-formula, Pareto dominance, and the lexicographic analysis [13, 14].

3.1.3 Selection Methods and Genetic Operators

After evaluating each individual, a selection method is employed for deciding which individuals will undergo reproduction, crossover and mutation. Some well-known selection methods are fitness-proportional selection, tournament selection, and linear ranking selection [2, 5].

In fitness-proportional selection, each individual i has probability p_i of being selected, which is proportional to its relative fitness, i.e., $p_i = f(i)/\sum_{j=1}^{I} f(j)$, where $f(.)$ is the fitness function and I is the total number of individuals in the GP population. In tournament selection, t individuals are selected at random from the population and the best individual from this t-group is selected. A common value for tournament size t is 2 [24]. In linear ranking selection, individuals are first ranked according to their fitness and then selected based on the value of their rank positions. This method overcomes the scaling problems of fitness-proportional assignment, e.g., premature convergence when few individuals with very high fitness values dominate the rest of the population.

Once the individuals have been selected, they undergo genetic operators such as reproduction (individual is cloned into the next generation), crossover (swapping of genetic material from two individuals, generating offspring) and mutation (modification of the genetic material of an individual).

There are many different ways to implement the crossover and mutation operators, and they also vary according to the individual encoding scheme. The *standard* EA crossover works by recombining the genetic material of two parent individuals, resulting in the generation of two children. For GAs, the most widely-used crossover strategies are one-point crossover, two-point crossover, and uniform crossover. In one-point crossover, the parents' linear strings are divided into two parts (according to the selected "point"), and the offspring is generated by using alternate parts of the parents. In two-point crossover, the rationale is the same but this time the string is divided into three, according to two distinct "points". Finally, uniform crossover is performed by swapping the parents genes in a per-gene fashion, given a swap probability. Figure 3.2 depicts these three crossover strategies usually adopted by GAs.

Mutation in linear genomes is usually performed in a per-gene basis, though there are exceptions such as the *flip bit* mutation, in which a binary vector is inverted (1 is changed to 0, and 0 is changed to 1). Usually, either a particular gene is selected to undergo mutation, or the entire genome is traversed and each gene has a probability of undergoing mutation. In integer and float genomes, operators such as *non-uniform*, *uniform*, and *Gaussian* mutation may be employed. Non-uniform mutation is employed with higher probability in the initial steps of evolution, to avoid the population from stagnating, and with lower probability in the later stages, only to

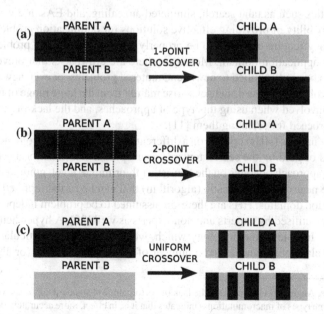

Fig. 3.2 Crossover strategies for linear strings: **a** one-point crossover; **b** two-point crossover; and **c** uniform crossover

fine-tune the individuals. Uniform mutation replaces the value of each gene with a random uniform value selected within the gene's range. Finally, Gaussian mutation adds a unit Gaussian-distributed random value to the chosen gene, and values outside the gene's boundaries are clipped.

In tree-based encoding schemes (GP), crossover occurs by swapping randomly chosen subtrees from the two parents. Regarding mutation, the standard approach is to replace a randomly selected individual's subtree by a randomly generated one. Even though crossover is the predominant operator in GP systems [3], there is a large controversy whether GP crossover is a constructive operator or a disruptive one. A constructive operator propagates good genetic material (good *building blocks*), increasing the quality of individuals across generations. A disruptive operator is a search strategy that mostly harms good solutions instead of improving them. In short, since GP crossover swaps subtrees from random positions in the tree, the larger the "good" building block contained in a tree, the higher the probability of it being disrupted after future crossover operations. This is the main reason for some researchers to claim that standard GP crossover is in fact a *macromutation* operator.[1] As a result of this controversy, new intelligent (context-aware, semantically-based, etc.) crossover operators are suggested for avoiding the negative effect of standard GP crossover (see, for instance, [28, 29, 41]).

3.2 Hyper-Heuristics

Metaheuristics such as tabu search, simulated annealing, and EAs, are well-known for their capability of providing effective solutions for optimisation problems. Nevertheless, they require expertise to be properly adopted for solving problems from a particular application domain. Moreover, there are drawbacks that prevent metaheuristics to be easily applied to newly encountered problems, or even new instances of known problems. These drawbacks arise mainly from the large range of parameter or choices involved when using this type of approaches, and the lack of guidance as to how to proceed for selecting them [11].

Hyper-heuristics (HHs) operate on a different level of generality from metaheuristics. Instead of guiding the search towards near-optimal solutions for a given problem, a HHs approach operates on the heuristic (algorithmic) level, guiding the search towards the near-optimal heuristic (algorithm) that can be further applied to different application domains. HHs are therefore assumed to be problem independent and can be easily utilised by experts and non-experts as well [35]. A hyper-heuristic can be seen as a high-level methodology which, when faced with a particular problem instance or class of instances, and a number of low-level heuristics (or algorithm's

---

[1] For instance, Angeline [1] argues that the lack of performance of standard GP crossover (in comparison to some types of macromutations) indicates that it is, in effect, more accurately described as a *population-limited macromutation operation* rather than an operation consistent with the building block hypothesis.

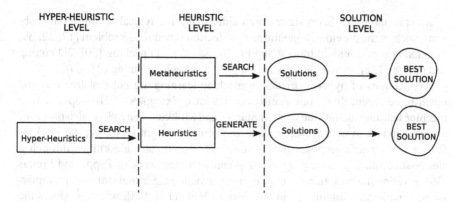

Fig. 3.3 Metaheuristics versus hyper-heuristics

design components), automatically designs a suitable combination of the provided components to effectively solve the respective problem(s) [11].

Figure 3.3 illustrates the different generality levels in which metaheuristics and hyper-heuristics work in. Note that whereas metaheuristics perform the search in the space of candidate solutions, hyper-heuristics perform the search in the space of candidate heuristics (algorithms), which in turn generate solutions for the problem at hand. To illustrate this rationale, let us compare two different evolutionary approaches in decision-tree induction. In the first approach, an EA is used to evolve the best decision tree for the *postoperative-patient* UCI data set [18]. In the second approach, an EA is used to evolve the best decision-tree algorithm to be further applied to medical data sets. Observe that, in the first approach, the EA works as a metaheuristic, because it searches for the best decision-tree to the *postoperative-patient data*. Therefore, the ultimate goal is to achieve an accurate decision tree for this particular problem. In the second approach, the EA works as a hyper-heuristic, because it searches for the best decision-tree algorithm, which in turn generates decision trees that can be applied to several different instances of medical applications. Note that the second approach is problem independent—instead of generating a decision tree that is only useful for classifying patients from the *postoperative-patient* data set, it generates a decision-tree algorithm that can be applied to several medical data sets, including the *postoperative patient* one.

The term hyper-heuristic is reasonably new, as it first appeared in a conference paper in 2000 [15], and in a journal paper in 2003 [8]. However, the methodology surrounding HHs are not new. It can be traced back to as early as the 1960s, when Fisher and Thompson [17] proposed the combination of scheduling rules (also known as priority or dispatching rules) for production scheduling, claiming that it would outperform any of the rules taken separately. This pioneering work was developed in a time when metaheuristics were not a mature research area. Even so, the proposed learning approach based on probabilistic learning resembles a stochastic local-search algorithm that operates in the space of scheduling rules sequences.

Most of the hyper-heuristic research aims at solving typical optimisation problems, such as the previously-mentioned production scheduling problem [17, 32, 39, 42], and also educational timetabling [9, 10, 33, 37], 1D packing [30], 2D cutting and packing [22], constraint satisfaction [40], and vehicle routing [21, 23].

Applications of hyper-heuristics in machine learning are not that frequent and much more recent than optimisation applications. Examples of HHs approach in machine learning include the work of Stanley and Miikkulainen [38], which proposes an evolutionary system for optimising the neural network topology; the work of Oltean [34], which proposes the evolution of evolutionary algorithms through a steady-state linear genetic programming approach; the work of Pappa and Freitas [36], covering the evolution of complete rule induction algorithms through grammar-based genetic programming; and the work of Vella et al. [43], which proposes the evolution of heuristic rules in order to select distinct split criteria in a decision-tree induction algorithm. In Chap. 4, we present a hyper-heuristic approach for evolving complete decision-tree induction algorithms.

3.3 Chapter Remarks

In this chapter, we presented the basic concepts regarding EAs and HHs. First, we showed the basic concepts of evolutionary computation from the perspective of genetic algorithms and genetic programming, the two most employed EAs in data mining and knowledge discovery [19]. In particular, we briefly described the main individual representations (encoding schemes) for the design of an EA, and also briefly discussed on the importance of fitness evaluation during the evolutionary cycle. Furthermore, we detailed the most common selection methods and genetic operators employed by researchers from this area.

In the second part of this chapter, we introduced the reader to *hyper-heuristics*. More specifically, we defined the differences between a metaheuristic approach and a hyper-heuristic approach, by providing an example that relates evolutionary algorithms and decision-tree induction. In addition, we presented the origins of hyper-heuristic research, and cited its main applications in the areas of optimisation and machine learning.

In Chap. 4, we present an approach for evolving full decision-tree induction algorithms through a hyper-heuristic evolutionary algorithm.

References

1. P.J. Angeline, Subtree crossover: building block engine or macromutation, in *Second Annual Conference on Genetic Programming*, pp. 9–17 (1997)
2. T. Back, Selective pressure in evolutionary algorithms: a characterization of selection mechanisms, in *IEEE Conference on Evolutionary Computation (CEC 1994)*, pp. 57–62 (1994)

3. W. Banzhaf et al., *Genetic Programming: An Introduction–On The Automatic Evolution of Computer Programs and its Applications* (Morgan Kaufmann Publishers Inc., San Francisco, 1998). ISBN: 1-55860-510-X
4. M.P. Basgalupp et al., Lexicographic multi-objective evolutionary induction of decision trees. Int. J. Bio-Inspir. Comput. **1**(1/2), 105–117 (2009)
5. T. Blickle, L. Thiele, *A Comparison of Selection Schemes Used in Evolutionary Algorithms* (Tech. Rep Swiss Federal Institute of Technology, Lausanne, 1995)
6. W. Bohm, A. Geyer-Schulz, *Foundations of Genetic Algorithms IV*, Exact uniform initialization for genetic programming (Morgan Kaufmann, San Francisco, 1996)
7. E. Burke, S. Gustafson, G. Kendall, Ramped half-n-half initialisation bias in GP, in *Genetic and Evolutionary Computation Conference (GECCO 2003)*, pp. 1800–1801 (2003)
8. E.K. Burke, G. Kendall, E. Soubeiga, A tabu-search hyperheuristic for timetabling and rostering. J. Heuristics **9**(6), 451–470 (2003)
9. E.K. Burke, S. Petrovic, R. Qu, Case-based heuristic selection for timetabling problems. J. Sched. **9**(2), 115–132 (2006). ISSN: 1094–6136
10. E.K. Burke et al., A graph-based hyper-heuristic for educational timetabling problems. Eur. J. Oper. Res. **176**(1), 177–192 (2007)
11. E.K. Burke et al., A survey of hyper-heuristics. Technical report Computer Science Technical Report No. NOTTCS-TR-SUB-0906241418-2747. School of Computer Science and Information Technology, University of Nottingham (2009)
12. K. Chellapilla, Evolving computer programs without subtree crossover. IEEE Trans. Evol. Comput. **1**(3), 209–216 (1997)
13. C. Coello, A comprehensive survey of evolutionary-based multiobjective optimization techniques. Knowl. Inf. Syst. **1**(3), 129–156 (1999)
14. C.A.C. Coello, G.B. Lamont, D.A.V. Veldhuizen, *Evolutionary Algorithms for Solving Multi-Objective Problems, Genetic and Evolutionary Computation* (Springer, New York, 2006)
15. P.I. Cowling, G. Kendall, E. Soubeiga, A hyperheuristic approach to scheduling a sales summit, in *Third International Conference on Practice and Theory of Automated Timetabling*. Springer, Berlin, pp. 176–190 (2001)
16. A.E. Eiben, J.E. Smith, *Introduction to Evolutionary Computing* (Springer, Berlin, 2003)
17. H. Fisher, G.L. Thompson, Probabilistic learning combinations of local job-shop scheduling rules, in *Industrial Scheduling*, ed. by J.F. Muth, G.L. Thompson (Prentice Hall, Upper Saddle River, 1963), pp. 225–251
18. A. Frank, A. Asuncion, *UCI Machine Learning Repository* (2010)
19. A.A. Freitas, Data Mining and Knowledge Discovery with Evolutionary Algorithms (Springer, New York, 2002). ISBN: 3540433317
20. A.A. Freitas, A review of evolutionary algorithms for data mining, in *Soft Computing for Knowledge Discovery and Data Mining*, ed. by O. Maimon, L. Rokach (Springer, Berlin, 2008), pp. 79–111. ISBN: 978-0-387-69935-6
21. P. Garrido, C. Castro, Stable solving of CVRPs using hyperheuristics, in *Proceedings of the 11th Annual conference on Genetic and evolutionary computation. GECCO'09*. Montreal, Québec, (Canada: ACM, 2009), pp. 255–262. ISBN: 978-1-60558-325-9
22. P. Garrido, M.-C. Riff, An evolutionary hyperheuristic to solve strip-packing problems, in *Proceedings of the 8th international conference on Intelligent data engineering and automated learning. IDEAL'07*. Springer, Birmingham, pp. 406–415 (2007)
23. P. Garrido, M.C. Riff, DVRP: a hard dynamic combinatorial optimisation problem tackled by an evolutionary hyper-heuristic. J. Heuristics **16**(6), 795–834 (2010). ISSN: 1381–1231
24. D.E. Goldberg, K. Deb, A comparative analysis of selection schemes used in genetic algorithms, in *Foundations of Genetic Algorithms*, ed. by G.J.E. Rawlins (Morgan Kaufmann, San Mateo, 1991), pp. 69–93
25. J.H. Holland, *Adaptation in Natural and Artificial Systems* (MIT Press, Cambridge, 1975)
26. J.R. Koza, *Genetic Programming: On the Programming of Computers by Means of Natural Selection* (MIT Press, Cambridge, 1992). ISBN: 0-262-11170-5

27. S. Luke, Two fast tree-creation algorithms for genetic programming. IEEE Trans. Evol. Comput. **4**(3), 274–283 (2000)
28. H. Majeed, C. Ryan, Using context-aware crossover to improve the performance of GP, in *8th Annual Conference on Genetic and Evolutionary Computation (GECCO'06)*. ACM, pp. 847–854 (2006)
29. H. Majeed, C. Ryan, A less destructive, context-aware crossover operator for GP, in *Lecture Notes in Computer Science*, ed. by P. Collet, et al. (Springer, Berlin, 2006), pp. 36–48
30. J.G. Marín-Blázquez, S. Schulenburg, A hyper-heuristic framework with XCS: learning to create novel problem-solving algorithms constructed from simpler algorithmic ingredients, in *Proceedings of the 2003–2005 International Conference on Learning Classifier Systems. IWLCS'03-05*. (Berlin: Springer, 2007), pp. 193–218. ISBN: 978-3-540-71230-5
31. M. Mitchell, *An Introduction to Genetic Algorithms* (MIT Press, Cambridge, 1998). ISBN: 0262631857
32. G. Ochoa et al., Dispatching rules for production scheduling: A hyper-heuristic landscape analysis, in *IEEE Congr. Evol. Comput.* pp. 1873–1880 (2009)
33. G. Ochoa, R. Qu, E.K. Burke. Analyzing the landscape of a graph based hyper-heuristic for timetabling problems, in *Proceedings of the 11th Annual conference on Genetic and Evolutionary Computation. GECCO'09*. Montreal, Québec, (Canada: ACM, 2009), pp. 341–348. ISBN: 978-1-60558-325-9
34. M. Oltean, Evolving evolutionary algorithms using linear genetic programming. Evol. Comput. **13**(3), 387–410 (2005)
35. E. Özcan, B. Bilgin, E.E. Korkmaz, A comprehensive analysis of hyper-heuristics. Intell. Data Anal. **12**(1), 3–23 (2008)
36. G.L. Pappa, A.A. Freitas, *Automating the Design of Data Mining Algorithms: An Evolutionary Computation Approach* (Springer Publishing Company Incorporated, New York, 2009)
37. N. Pillay, An analysis of representations for hyper-heuristics for the uncapacitated examination timetabling problem in a genetic programming system, in *Proceedings of the 2008 Annual Research Conference of the South African Institute of Computer Scientists and Information Technologists on IT Research in Developing Countries: Riding the Wave of Technology. SAICSIT'08*. Wilderness, (South Africa: ACM, 2008), pp. 188–192. ISBN: 978-1-60558-286-3
38. K.O. Stanley, R. Miikkulainen, Evolving neural networks through augmenting topologies. Evol. Comput. **10**(2), 99–127 (2002). ISSN: 1063–6560
39. R.H. Storer, S.D. Wu, R. Vaccari, New search spaces for sequencing problems with application to job shop scheduling. Manag. Sci. **38**(10), 1495–1509 (1992)
40. H. Terashima-Marín et al., Generalized hyper-heuristics for solving 2D regular and irregular packing problems. Ann. Oper. Res. **179**(1), 369–392 (2010)
41. N. Uy et al., in *Lecture Notes in Computer Science*, Semantic similarity based crossover in GP: The case for real-valued function regression, ed. by P. Collet, et al. (Springer, Berlin, 2010), pp. 170–181
42. J.A. Vázquez-Rodríguez, S. Petrovic, A new dispatching rule based genetic algorithm for the multi-objective job shop problem. J. Heuristics **16**(6), 771–793 (2010)
43. A. Vella, D. Corne, C. Murphy, Hyper-heuristic decision tree induction, in *World Congress on Nature and Biologically Inspired Computing*, pp. 409–414 (2010)
44. T. Weise, Global Optimization Algorithms—Theory and Application. en. Second. Online available at http://www.it-weise.de/. Accessed in Sept 2009. Self-Published, (2009)
45. J.R. Woodward, GA or GP? That is not the question, in *IEEE Congress on Evol. Comput. (CEC 2003)*. pp. 1056–1063 (2003)

Chapter 4
HEAD-DT: Automatic Design of Decision-Tree Algorithms

Abstract As presented in Chap. 2, for the past 40 years researchers have attempted to improve decision-tree induction algorithms, either by proposing new splitting criteria for internal nodes, by investigating pruning strategies for avoiding overfitting, by testing new approaches for dealing with missing values, or even by searching for alternatives to the top-down greedy induction. Each new decision-tree induction algorithm presents some (or many) of these strategies, which are chosen in order to maximize performance in empirical analyses. Nevertheless, the number of different strategies for the several components of a decision-tree algorithm is so vast after these 40 years of research that it would be impracticable for a human being to test all possibilities with the purpose of achieving the best performance in a given data set (or in a set of data sets). Hence, we pose two questions for researchers in the area: "is it possible to automate the design of decision-tree induction algorithms?", and, if so, "how can we automate the design of a decision-tree induction algorithm?" The answer for these questions arose with the pioneering work of Pappa and Freitas [30], which proposed the automatic design of rule induction algorithms through an evolutionary algorithm. The authors proposed the use of a grammar-based GP algorithm for building and evolving individuals which are, in fact, rule induction algorithms. That approach successfully employs EAs to evolve a generic rule induction algorithm, which can then be applied to solve many different classification problems, instead of evolving a specific set of rules tailored to a particular data set. As presented in Chap. 3, in the area of optimisation this type of approach is named hyper-heuristics (HHs) [5, 6]. HHs are search methods for automatically selecting and combining simpler heuristics, resulting in a generic heuristic that is used to solve any instance of a given optimisation problem. For instance, a HH can generate a generic heuristic for solving any instance of the timetabling problem (i.e., allocation of any number of resources subject to any set of constraints in any schedule configuration) whilst a conventional EA would just evolve a solution to one particular instance of the timetabling problem (i.e., a predefined set of resources and constraints in a given schedule configuration). In this chapter, we present a hyper-heuristic strategy for automatically designing decision-tree induction algorithms, namely HEAD-DT (Hyper-Heuristic Evolutionary Algorithm for Automatically Designing Decision-Tree Algorithms). Section 4.1 introduces HEAD-DT and its evolutionary scheme. Section 4.2 presents the individual representation adopted by HEAD-DT to evolve decision-tree algorithms, as

© The Author(s) 2015

R.C. Barros et al., *Automatic Design of Decision-Tree Induction Algorithms*,
SpringerBriefs in Computer Science, DOI 10.1007/978-3-319-14231-9_4

well as information regarding each individual's gene. Section 4.3 shows the evolutionary cycle of HEAD-DT, detailing its genetic operators. Section 4.4 depicts the fitness evaluation process in HEAD-DT, and introduces two possible frameworks for executing HEAD-DT. Section 4.5 computes the total size of the search space that HEAD-DT is capable of traversing, whereas Sect. 4.6 discusses related work.

Keywords Automatic design · Hyper-heuristic decision-tree induction · HEAD-DT

4.1 Introduction

According to the definition by Burke et al. [7], "a hyper-heuristic is an automated methodology for selecting or generating heuristics to solve hard computational search problems". Hyper-heuristics can automatically generate new heuristics suited to a given problem or class of problems. This is carried out by combining components or building-blocks of human-designed heuristics. The motivation behind hyper-heuristics is to raise the level of generality at which search methodologies can operate. In the context of decision trees, instead of searching through an EA for the best decision tree to a given problem (regular metaheuristic approach, e.g., [1, 2]), the generality level is raised by searching for the best decision-tree induction algorithm that may be applied to several different problems (hyper-heuristic approach).

HEAD-DT (Hyper-Heuristic Evolutionary Algorithm for Automatically Designing Decision-Tree Algorithms) can be seen as a regular generational EA in which individuals are collections of building blocks of top-down decision-tree induction algorithms. Typical operators from EAs are employed, such as tournament selection, mutually-exclusive genetic operators (reproduction, crossover, and mutation) and a typical stopping criterion that halts evolution after a predefined number of generations. The evolution of individuals in HEAD-DT follows the scheme presented in Fig. 4.1.

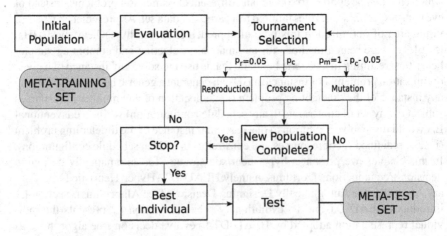

Fig. 4.1 HEAD-DT's evolutionary scheme

4.2 Individual Representation

Each individual in HEAD-DT is encoded as an integer vector, as depicted in Fig. 4.2, and each gene has a different range of supported values. We divided the genes into four categories, representing the major building blocks (design components) of a decision-tree induction algorithm:

- split genes;
- stopping criteria genes;
- missing values genes;
- pruning genes.

4.2.1 Split Genes

The linear genome that encodes individuals in HEAD-DT holds two genes for the **split** component of decision trees. These genes represent the design component that is responsible for selecting the attribute to split the data in the current node of the decision tree. Based on the selected attribute, a decision rule is generated for filtering the input data in subsets, and the process continues recursively.

To model this design component, we make use of two different genes. The first one, **criterion**, is an integer that indexes one of the 15 splitting criteria that are implemented in HEAD-DT (see Table 4.1). The most successful criteria are based on Shannon's entropy [36], a concept well-known in information theory. Entropy is a unique function that satisfies the four axioms of uncertainty. It represents the average amount of information when coding each class into a codeword with ideal length according to its probability. Examples of splitting criteria based on entropy are *global mutual information* (GMI) [18] and *information gain* [31]. The latter is employed by Quinlan in his ID3 system [31]. However, Quinlan points out that information gain

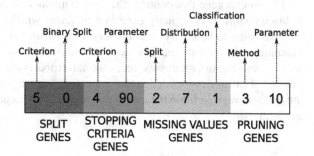

Fig. 4.2 Linear-genome for evolving decision-tree algorithms

Table 4.1 Split criteria
implemented in HEAD-DT

Criterion	References
Information gain	[31]
Gini index	[4]
Global mutual information	[18]
G statistic	[26]
Mantáras criterion	[24]
Hypergeometric distribution	[25]
Chandra-Varghese criterion	[9]
DCSM	[10]
χ^2	[27]
Mean posterior improvement	[37]
Normalized gain	[21]
Orthogonal criterion	[15]
Twoing	[4]
CAIR	[11]
Gain ratio	[35]

is biased towards attributes with many values, and thus proposes a solution named
gain ratio [35]. Gain ratio normalizes the information gain by the entropy of the
attribute being tested. Several variations of the gain ratio have been proposed, such
as the normalized gain [21].

Alternatives to entropy-based criteria are the class of distance-based measures.
These criteria evaluate separability, divergency, or discrimination between classes.
Examples are the Gini index [4], the twoing criterion [4], the orthogonality criterion
[15], among several others. We also implemented as options for HEAD-DT lesser-
known criteria such as CAIR [11] and mean posterior improvement [37], as well as
the more recent Chandra-Varghese [9] and DCSM [10], to enhance the diversity of
options for generating splits in a decision tree.

The second gene that controls the split component of a decision-tree algorithm
is **binary split**. It is a binary gene that indicates whether the splits of a decision
tree are going to be binary or multi-way. In a binary tree, every split has only two
outcomes, which means that nominal attributes with many categories are aggregated
into two subsets. In a multi-way tree, nominal attributes are divided according to their
number of categories—one edge (outcome) for each category. In both cases, numeric
attributes always partition the tree in two subsets, represented by tests $att \leq \Delta$ and
$att > \Delta$.

4.2.2 Stopping Criteria Genes

The top-down induction of decision trees is recursive and it continues until a stopping criterion (also known as *pre-pruning*) is satisfied. The linear genome in HEAD-DT holds two genes for representing this design component: **criterion** and **parameter**.

The first gene, **criterion**, selects among the following five different strategies for stopping the tree growth:

1. Reaching class homogeneity: when every instance that reaches a given node belong to the same class, there is no reason to split this node any further. This strategy can be the only single stopping criterion, or it can be combined with the next four strategies;
2. Reaching the maximum tree depth: a parameter *tree depth* can be specified to avoid deep trees. Range: [2, 10] levels;
3. Reaching the minimum number of instances for a non-terminal node: a parameter *minimum number of instances for a non-terminal node* can be specified to avoid/alleviate the data fragmentation problem in decision trees. Range: [1, 20] instances;
4. Reaching the minimum percentage of instances for a non-terminal node: same as above, but instead of the actual number of instances, a percentage of instances is defined. The parameter is thus relative to the total number of instances in the training set. Range: [1 %, 10 %] the total number of instances;
5. Reaching an accuracy threshold within a node: a parameter *accuracy reached* can be specified for halting the growth of the tree when the accuracy within a node (majority of instances) has reached a given threshold. Range: {70 %, 75 %, 80 %, 85 %, 90 %, 95 %, 99 %} accuracy.

Gene **parameter** dynamically adjusts a value in the range [0, 100] to the corresponding strategy. For example, if the strategy selected by gene **criterion** is *reaching the maximum tree depth*, the following mapping function is executed:

$$param = (value \bmod 9) + 2 \tag{4.1}$$

This function maps from [0, 100] (variable *value*) to [2, 10] (variable *param*), which is the desired range of values for the parameter of strategy *reaching the maximum tree depth*. Similar mapping functions are executed dynamically to adjust the ranges of gene **parameter**.

4.2.3 Missing Values Genes

The next design component of decision trees that is represented in the linear genome of HEAD-DT is the *missing value treatment*. Missing values may be an issue during tree induction and also during classification. We make use of three genes to represent missing values strategies in different moments of the induction/deduction process. During tree induction, there are two moments in which we need to deal with missing values: splitting criterion evaluation (**split gene**), and instances distribution (**distribution gene**). During tree deduction (classification), we may also have to deal with missing values in the test set (**classification gene**).

During the split criterion evaluation in node t based on attribute a_i, we implemented the following strategies:

- Ignore all instances whose value of a_i is missing [4, 17];
- Imputation of missing values with either the mode (nominal attributes) or the mean/median (numeric attributes) of all instances in t [12];
- Weight the splitting criterion value (calculated in node t with regard to a_i) by the proportion of missing values [34];
- Imputation of missing values with either the mode (nominal attributes) or the mean/median (numeric attributes) of all instances in t whose class attribute is the same of the instance whose a_i value is being imputed.

For deciding which child node training instance x_j should go to, considering a split in node t over a_i, we adopted the options:

- Ignore instance x_j [31];
- Treat instance x_j as if it has the most common value of a_i (mode or mean), regardless of the class [34];
- Treat instance x_j as if it has the most common value of a_i (mode or mean) considering the instances that belong to the same class than x_j;
- Assign instance x_j to all partitions [17];
- Assign instance x_j to the partition with largest number of instances [34];
- Weight instance x_j according to the partition probability [22, 35];
- Assign instance x_j to the most probable partition, considering the class of x_j [23].

Finally, for classifying an unseen test instance x_j, considering a split in node t over a_i, we used the strategies:

- Explore all branches of t combining the results [32];
- Take the route to the most probable partition (largest subset);
- Halt the classification process and assign instance x_j to the majority class of node t [34].

4.2.4 Pruning Genes

Pruning was originally conceived as a strategy for tolerating noisy data, though it was found to improve decision tree accuracy in many noisy data sets [4, 31, 33]. It has now become an important part of greedy top-down decision-tree induction algorithms. HEAD-DT holds two genes for this design component. The first gene, **method**, indexes one of the five well-known approaches for pruning a decision tree

Table 4.2 Pruning methods implemented in HEAD-DT

Method	References
Reduced-error pruning	[33]
Pessimistic error pruning	[33]
Minimum error pruning	[8, 28]
Cost-complexity pruning	[4]
Error-based pruning	[35]

presented in Table 4.2, and also the option of not pruning at all. The second gene, **parameter**, is in the range [0, 100] and its value is again dynamically mapped by a function according to the selected pruning method.

(1) Reduced-error pruning (REP) is a conceptually simple strategy proposed by Quinlan [33]. It uses a pruning set (a part of the training set) to evaluate the goodness of a given subtree from T. The idea is to evaluate each non-terminal node t with regard to the classification error in the pruning set. If such an error decreases when we replace the subtree $T^{(t)}$ rooted on t by a leaf node, then $T^{(t)}$ must be pruned. Quinlan imposes a constraint: a node t cannot be pruned if it contains a subtree that yields a lower classification error in the pruning set. The practical consequence of this constraint is that REP should be performed in a bottom-up fashion. The REP pruned tree T' presents an interesting optimality property: it is the smallest most accurate tree resulting from pruning original tree T [33]. Besides this optimality property, another advantage of REP is its linear complexity, since each node is visited only once in T. A clear disadvantage is the need of using a pruning set, which means one has to divide the original training set, resulting in less instances to grow the tree. This disadvantage is particularly serious for small data sets. For REP, the parameter gene is regarding the percentage of training data to be used in the pruning set (varying within the interval [10 %, 50 %]).

(2) Also proposed by Quinlan [33], the pessimistic error pruning (PEP) uses the training set for both growing and pruning the tree. The apparent error rate, i.e., the error rate calculated over the training set, is optimistically biased and cannot be used to decide whether pruning should be performed or not. Quinlan thus proposes adjusting the apparent error according to the continuity correction for the binomial distribution in order to provide a more realistic error rate. PEP is computed in a top-down fashion, and if a given node t is pruned, its descendants are not examined, which makes this pruning strategy quite efficient in terms of computational effort. Esposito et al. [14] point out that the introduction of the continuity correction in the estimation of the error rate has no theoretical justification, since it was never applied to correct over-optimistic estimates of error rates in statistics. For PEP, the parameter gene is the number of standard errors (SEs) to adjust the apparent error, in the set {0.5, 1, 1.5, 2}.

(3) Originally proposed by Niblett and Bratko [28] and further extended in [8], minimum error pruning (MEP) is a bottom-up approach that seeks to minimize the *expected error rate* for unseen cases. It uses an ad-hoc parameter m for controlling

the level of pruning. Usually, the higher the value of m, the more severe the pruning. Cestnik and Bratko [8] suggest that a domain expert should set m according to the level of noise in the data. Alternatively, a set of trees pruned with different values of m could be offered to the domain expert, so he/she can choose the best one according to his/her experience. For MEP, the parameter gene comprises variable m, which may range within [0, 100].

(4) Cost-complexity pruning (CCP) is the post-pruning strategy adopted by the CART system [4]. It consists of two steps: (i) generate a sequence of increasingly smaller trees, beginning with T and ending with the root node of T, by successively pruning the subtree yielding the lowest *cost complexity*, in a bottom-up fashion; (ii) choose the best tree among the sequence based on its relative size and accuracy (either on a pruning set, or provided by a cross-validation procedure in the training set). The idea within step 1 is that pruned tree T_{i+1} is obtained by pruning the subtrees that show the lowest increase in the apparent error (error in the training set) per pruned leaf. Regarding step 2, CCP chooses the smallest tree whose error (either on the pruning set or on cross-validation) is not more than one standard error (SE) greater than the lowest error observed in the sequence of trees. For CCP, there are two parameters that need to be set: the number of SEs (in the same range than PEP) and the pruning set size (in the same range than REP).

(5) Error-based pruning (EBP) was proposed by Quinlan and it is implemented as the default pruning strategy of C4.5 [35]. It is an improvement over PEP, based on a far more pessimistic estimate of the expected error. Unlike PEP, EBP performs a bottom-up search, and it carries out not only the replacement of non-terminal nodes by leaves but also *grafting* of subtree $T^{(t)}$ onto the place of parent t. For deciding whether to replace a non-terminal node by a leaf (subtree replacement), to graft a subtree onto the place of its parent (subtree raising) or not to prune at all, a pessimistic estimate of the expected error is calculated by using an upper confidence bound. An advantage of EBP is the new *grafting* operation that allows pruning useless branches without ignoring interesting lower branches. A drawback of the method is the presence of an ad-hoc parameter, *CF*. Smaller values of *CF* result in more pruning. For EBP, the parameter gene comprises variable *CF*, which may vary within [1 %, 50 %].

4.2.5 Example of Algorithm Evolved by HEAD-DT

The linear genome of an individual in HEAD-DT is formed by the building blocks described in the earlier sections: (*split criterion, split type, stopping criterion, stopping parameter, pruning strategy, pruning parameter, mv split, mv distribution, mv classification.*) One possible individual encoded by that linear string is [4, 1, 2, 77, 3, 91, 2, 5, 1], which accounts for Algorithm 1.

Algorithm 1 Example of a decision-tree algorithm automatically designed by HEAD-DT.

1: Recursively split nodes with the G statistics criterion;
2: Create one edge for each category in a nominal split;
3: Perform step 1 until class-homogeneity or the maximum tree depth of 7 levels ((77 mod 9) + 2) is reached;
4: Perform MEP pruning with $m = 91$;
   When dealing with missing values:
5: Distribute missing-valued instances to the partition with the largest number of instances;
6: Distribute missing values by assigning the instance to all partitions;
7: For classifying an instance with missing values, explore all branches and combine the results.

4.3 Evolution

The first step of HEAD-DT is the generation of the initial population, in which a population of 100 individuals (default value) is randomly generated (generation of random numbers within the genes' acceptable range of values). Next, the population's fitness is evaluated based on the data sets that belong to the meta-training set. The individuals then participate of a pairwise tournament selection procedure ($t = 2$ is the default parameter) for defining those that will undergo the genetic operators. Individuals may participate in either uniform crossover, random uniform gene mutation, or reproduction, the three mutually-exclusive genetic operators employed in HEAD-DT (see Fig. 4.3).

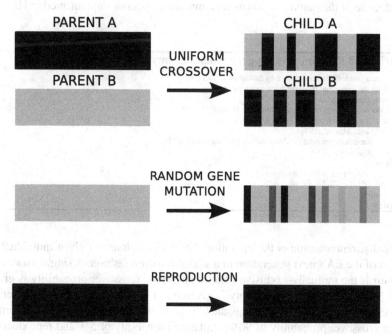

Fig. 4.3 HEAD-DT's genetic operators

The uniform crossover is guided by a swap probability p_s (default value $= 0.5$) that ultimately indicates whether a child's gene should come from parent A or from parent B. Algorithm 2 depicts the pseudocode of the uniform crossover operator implemented in HEAD-DT.

Algorithm 2 Uniform crossover employed by HEAD-DT.

1: Let A and B be two parents chosen by tournament selection;
2: Let C and D be the two resulting offspring;
3: **for** each gene g in genome **do**
4:      Choose a uniform random real number u from [0,1];
5:      **if** $u \leq p_s$ **then**
6:          //swap genes
7:          $C[g] = B[g]$;
8:          $D[g] = A[g]$;
9:      **else**
10:          //do not swap
11:          $C[g] = A[g]$;
12:          $D[g] = B[g]$;
13:      **end if**
14: **end for**

The mutation operator implemented in HEAD-DT is the random uniform gene mutation. It is guided by a replacement probability p_{rep} (default value $= 0.5$), which dictates whether or not a gene's value should be replaced by a randomly generated value within the accepted range of the respective gene. Algorithm 3 depicts the pseudocode of the random uniform gene mutation operator implemented in HEAD-DT.

Algorithm 3 Random uniform gene mutation employed by HEAD-DT.

1: Let A be a single individual chosen by tournament selection;
2: Let B be the individual resulting from mutating A;
3: **for** each gene g in genome **do**
4:      Choose a uniform random real number u from [0,1];
5:      **if** $u \leq p_{rep}$ **then**
6:          //mutate gene
7:          Randomly generate a value v within the range accepted by g;
8:          $B[g] = v$;
9:      **else**
10:          //do not mutate gene
11:          $B[g] = A[g]$
12:      **end if**
13: **end for**

Finally, reproduction is the operation that copies (clones) a given individual to be part of the EA's next generation in a straightforward fashion. A single parameter p controls the mutually-exclusive genetic operators: crossover probability is given by p, whereas mutation probability is given by $(1 - p) - 0.05$, and reproduction probability is fixed in 0.05. For instance, if $p = 0.9$, then HEAD-DT is executed with a crossover probability of 90 %, mutation probability of 5 % and reproduction probability of 5 %. HEAD-DT employs an elitist strategy, in which the best e % individuals are kept from one generation to the next ($e = 5$ % of the population is

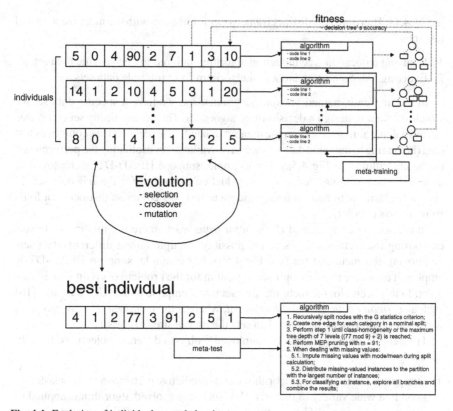

Fig. 4.4 Evolution of individuals encoded as integer vectors

the default parameter). Evolution ends after a predefined number of generations is achieved (100 generations is the default value), and the best individual returned by HEAD-DT is then executed over the meta-test set, so its performance in unseen data can be properly assessed.

Figure 4.4 presents an example of how linear genomes are decoded into algorithms, and how they participate of the evolutionary cycle. For decoding the individuals, the building blocks (indexed components and their respective parameters) are identified, and this information is passed to a skeleton decision-tree induction class, filling the gaps with the selected building blocks.

4.4 Fitness Evaluation

During the fitness evaluation, HEAD-DT employs a *meta-training set* for assessing the quality of each individual throughout evolution. A *meta-test set* is used for assessing the quality of the evolved decision-tree induction algorithm (the best individual in Fig. 4.1). There are two distinct frameworks for dealing with the meta-training and test sets:

1. Evolving a decision-tree induction algorithm tailored to one specific data set.
2. Evolving a decision-tree induction algorithm from multiple data sets.

In the first case, named *the specific framework*, we have a specific data set for which we want to design a decision-tree algorithm. The meta-training set comprises the available training data from the data set at hand. The meta-test set comprises test data (belonging to the same data set) we have available for evaluating the performance of the algorithm (see Fig. 4.5a). For example, suppose HEAD-DT is employed to evolve the near-optimal decision-tree induction algorithm for the *iris* data set. In such a scenario, both meta-training and meta-test sets comprise distinct data folds from the *iris* data set.

In the second case, named *the general framework*, there are multiple data sets composing the meta-training set, and possibly multiple (albeit different) data sets comprising the meta-test set (see Fig. 4.5b). For example, suppose HEAD-DT is employed to evolve the near-optimal algorithm for the problem of credit risk assessment. In this scenario, the meta-training set may comprise public UCI data sets [16] such as *german credit* and *credit approval*, whereas the meta-test set may comprise particular credit risk assessment data sets the user desires to classify.

The general framework can be employed with two different objectives, broadly speaking:

1. Designing a decision-tree algorithm whose predictive performance is consistently good in a wide variety of data sets. For such, the evolved algorithm is applied to data sets with very different structural characteristics and/or from very distinct application domains. In this scenario, the user chooses distinct data sets to be part of the meta-training set, in the hope that evolution will be capable of generating an algorithm that performs well in a wide range of data sets. Pappa [29] calls this strategy *"evolving robust algorithms"*;
2. Designing a decision-tree algorithm that is tailored to a particular application domain or to a specific statistical profile. In this scenario, the meta-training set comprises data sets that share *similarities*, and so the evolved decision-tree algorithm is specialized in solving a specific type of problem. Unlike the previous

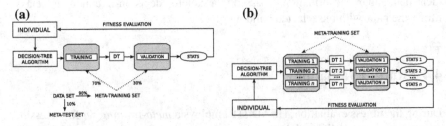

Fig. 4.5 Fitness evaluation schemes. **a** Fitness evaluation from one data set in the meta-training set. **b** Fitness evaluation from multiple data sets in the meta-training set

strategy, in this case we have to define a *similarity* criterion for creating specialized algorithms. We highlight the following *similarity* criteria: (i) choosing
data sets that share the same *application domain* (e.g., gene expression data); (ii)
choosing data sets with *provenance resemblance* (e.g., data sets generated from
data collected by a specific sensor); and (iii) choosing data sets with *structural
resemblance* (e.g., data sets with statistically-similar features and/or with similar
geometrical complexity [19, 20]).

In Fig. 4.5b, we can observe how the fitness evaluation of a decision-tree induction
algorithm evolved from multiple data sets occurs. First, a given individual is mapped
into its corresponding decision-tree algorithm. Afterwards, each data set that belongs
to the meta-training set is divided into training and validation—typical values are
70 % for training and 30 % for validation [39]. The term "validation set" is used in
here instead of "test set" to avoid confusion with the meta-test set, and also due to the
fact that we are using the "knowledge" within these sets to reach for a better solution
(the same cannot be done with test sets, which are exclusively used for assessing the
performance of an algorithm).

After dividing each data set from the meta-training set into "training" and "validation", a decision tree is induced for each training set available. For evaluating the
performance of these decision trees, we use the corresponding validation sets. Statistics regarding the performance of each decision tree are recorded (e.g., accuracy,
F-Measure, precision, recall, total number of nodes/leaves, etc.), and can be used
individually or combined as the fitness function of HEAD-DT. The simple average is
probably the most intuitive way of combining the values per data set, but other possible solutions are the median of the values, or their harmonic mean. Depending on
the data sets used in the meta-training set, the user may decide to give greater weight
of importance to a more difficult-to-solve data set than to an easier one, and hence a
weighted scheme may be a good solution when combining the data set values. Some
of these possibilities are discussed in Chap. 6.

A typical fitness function would be the average F-Measure of the decision trees
generated by a given individual for each data set from the meta-training set. F-
Measure (also known as F-score or F1 score) is the harmonic mean of precision and
recall:

$$precision = \frac{tp}{tp + fp} \tag{4.2}$$

$$recall = \frac{tp}{tp + fn} \tag{4.3}$$

$$fmeasure = 2 \times \frac{precision \times recall}{precision + recall} \tag{4.4}$$

$$Fitness = \frac{1}{n} \sum_{i=1}^{n} fmeasure_i \tag{4.5}$$

where tp (tn) is the number of true positives (negatives), fp (fn) is the number of false positives (negatives), $fmeasure_i$ is the F-Measure obtained in data set i and n is the total number of data sets in the meta-training set.

This formulation assumes that the classification problem at hand is binary, i.e., composed by two classes: positive and negative. Nevertheless, it can be trivially extended to multi-class problems. For instance, we can compute the measure for each class—assuming each class to be the "positive" class in turn—and (weight-)average the per-class measures. Having in mind that we would like HEAD-DT to perform well in both balanced and imbalanced data sets, we believe that the average F-Measure is a more suitable fitness function than the average accuracy.

4.5 Search Space

To compute the search space reached by HEAD-DT, consider the linear genome presented in Sect. 4.2: (*split criterion, split type, stopping criterion, stopping parameter, pruning strategy, pruning parameter, mv split, mv distribution, mv classification*). There are 15 types of split criteria, 2 possible split types, 4 types of missing-value strategies during split computation, 7 types of missing-value strategies during training data distribution, and 3 types of missing-value strategies during classification. Hence, there are $15 \times 2 \times 4 \times 7 \times 3 = 2{,}520$ possible different algorithms.

Now, let us analyse the combination of stopping criteria and their parameters. There is the possibility of splitting until class homogeneity is achieved, and no parameter is needed (thus, 1 possible algorithm). There are 9 possible parameters when the tree is grown until a maximum depth, and 20 when reaching a minimum number of instances. Furthermore, there are 10 possible parameter values when reaching a minimum percentage of instances and 7 when reaching an accuracy threshold. Hence, there are $1 + 9 + 20 + 10 + 7 = 47$ possible algorithms just by varying the stopping criteria component.

Next, let us analyse the combination of pruning methods and their parameters. REP pruning parameter may take up to 5 different values, whereas PEP pruning may take up to 4. MEP can take up to 101 values, and EBP up to 50. Finally, CCP takes up to 4 values for its first parameter and up to 5 values for its second. Therefore, there are $5 + 4 + 101 + (4 \times 5) + 50 = 180$ possible algorithms by just varying the pruning component.

If we combine all the previously mentioned values, HEAD-DT currently searches in the space of $2{,}520 \times 47 \times 180 = 21{,}319{,}200$ algorithms. Now, just for the sake of argument, suppose a single decision-tree induction algorithm takes about 10 s to produce a decision tree for a given (small) data set for which we want the best possible algorithm. If we were to try all possible algorithms in a brute-force approach, we would take 59,220 h to find out the best possible configuration for that data set. That means ≈2,467 days or 6.75 years just to find out the best decision-tree algorithm for a single (small) data set. HEAD-DT would take, in the worst case, 100,000 s—10,000 individuals (100 individuals per generation, 100 generations) times 10 s. Thus HEAD-DT would take about 1,666 min (27.7 h) to compute the (near-)optimal algorithm for that same data set, i.e., it is ≈2,138 times faster than the brute-force approach. In practice, this number is much smaller considering that individuals are not re-evaluated if not changed, and HEAD-DT implements reproduction and elitism.

Of course there are no theoretic guarantees that the (near-)optimal algorithm found by HEAD-DT within these 27.7 h is going to be the same global optimal algorithm provided by the brute-force approach after practically 7 years of computation, but its use is justified by the time saved during the process.

4.6 Related Work

The literature in EAs for decision-tree induction is very rich (see, for instance, [3]). However, the research community is still concerned with the evolution of decision trees for particular data sets instead of evolving full decision-tree induction algorithms.

To the best of our knowledge, no work to date attempts to automatically design full decision-tree induction algorithms. The most related approach to the one presented in this book is HHDT (Hyper-Heuristic Decision Tree) [38]. It proposes an EA for evolving heuristic rules in order to determine the best splitting criterion to be used in non-terminal nodes. It is based on the degree of entropy of the data set attributes. For instance, it evolves rules such as *IF* ($x\%\geq high$) and ($y\%<low$) *THEN use heuristic A*, where x and y are percentages ranging within [0, 100], and high and low are threshold entropy values. This rule has the following interpretation: if $x\%$ of the attributes have entropy values greater or equal than threshold *high*, and if $y\%$ of the attributes have entropy values below threshold *low*, then make use of heuristic A for choosing the attribute that splits the current node. Whilst HHDT is a first step to automate the design of decision-tree induction algorithms, it evolves a single component of the algorithm (the choice of splitting criterion), and thus should be further extended for being able to generate full decision-tree induction algorithms, which is the case of HEAD-DT.

Another slightly related approach is the one presented by Delibasic et al. [13]. The authors propose a framework for combining decision-tree components, and test 80 different combination of design components on 15 benchmark data sets. This approach is not a hyper-heuristic, since it does not present an heuristic to choose among different heuristics. It simply selects a fixed number of component combinations and test them all against traditional decision-tree algorithms (C4.5, CART, ID3 and CHAID). We believe that employing EAs to evolve decision-tree algorithms is a more robust strategy, since we can search for solutions in a much larger search space (21 million possible algorithms in HEAD-DT, against 80 different algorithms in the work of Delibasic et al. [13]).

Finally, the work of Pappa and Freitas [30] proposes a grammar-based genetic programming approach (GGP) for evolving full rule induction algorithms. The results showed that GGP could generate rule induction algorithms different from those already proposed in the literature, and with competitive predictive performance.

4.7 Chapter Remarks

In this chapter, we presented HEAD-DT, a hyper-heuristic evolutionary algorithm that automatically designs top-down decision-tree induction algorithms. The latter have been manually improved for the last 40 years, resulting in a large number of approaches for each of their design components. Since the human manual approach for testing all possible modifications in the design components of decision-tree algorithms would be unfeasible, we believe the evolutionary search of HEAD-DT constitutes a robust and efficient solution for the problem.

HEAD-DT evolves individuals encoded as integer vectors (linear genome). Each gene in the vector is an index to a design component or the value of its corresponding parameter. Individuals are decoded by associating each integer to a design component, and by mapping values ranging within [0, 100] to values in the correct range according to the specified component. The initial population of 100 individuals evolve for 100 generations, in which individuals are chosen by a pairwise tournament selection strategy to participate of mutually-exclusive genetic operators such as uniform crossover, random uniform gene mutation, and reproduction.

HEAD-DT may operate under two distinct frameworks: (i) evolving a decision-tree induction algorithm tailored to one specific data set; and (ii) evolving a decision-tree induction algorithm from multiple data sets. In the first framework, the goal is to generate a decision-tree algorithm that excels at a single data set (both meta-training and meta-test sets comprise data from the same data set). In the second framework, there are several distinct objectives that can be achieved, like generating a decision-tree algorithm tailored to a particular application domain (say gene expression data sets or financial data sets), or generating a decision-tree algorithm that is robust across several different data sets (a good "all-around" algorithm).

Regardless of the framework being employed, HEAD-DT is capable of searching in a space of more than 21 million algorithms. In the next chapter, we present several experiments for evaluating HEAD-DT's performance under the two proposed frameworks. Moreover, we comment on the cost-effectiveness of automated algorithm design in contrast to the manual design, and we show that the genetic search performed by HEAD-DT is significantly better than a random search in the space of 21 million decision-tree induction algorithms.

References

1. R.C. Barros, D.D. Ruiz, M.P. Basgalupp, Evolutionary model trees for handling continuous classes in machine learning. Inf. Sci. **181**, 954–971 (2011)
2. R.C. Barros et al., Towards the automatic design of decision tree induction algorithm, in *13th Annual Conference Companion on Genetic and Evolutionary Computation* (GECCO 2011). pp. 567–574 (2011)
3. R.C. Barros et al., A survey of evolutionary algorithms for decision-tree induction. IEEE Trans. Syst., Man, Cybern., Part C: Appl. Rev. **42**(3), 291–312 (2012)
4. L. Breiman et al., *Classification and Regression Trees* (Wadsworth, Belmont, 1984)
5. E. Burke, S. Petrovic, Recent research directions in automated timetabling. Eur. J. Oper. Res. **140**(2), 266–280 (2002)
6. E.K. Burke, G. Kendall, E. Soubeiga, A tabu-search hyperheuristic for timetabling and rostering. J. Heuristics **9**(6), 451–470 (2003)
7. E.K. Burke et al., A Classification of Hyper-heuristics Approaches, in *Handbook of Metaheuristics*, 2nd edn., International Series in Operations Research & Management Science, ed. by M. Gendreau, J.-Y. Potvin (Springer, Berlin, 2010), pp. 449–468
8. B. Cestnik, I. Bratko, *On Estimating Probabilities in Tree Pruning*, Machine learning-EWSL-91. Vol. 482. Lecture Notes in Computer Science (Springer, Berlin, 1991)
9. B. Chandra, P.P. Varghese, Moving towards efficient decision tree construction. Inf. Sci. **179**(8), 1059–1069 (2009)
10. B. Chandra, R. Kothari, P. Paul, A new node splitting measure for decision tree construction. Pattern Recognit. **43**(8), 2725–2731 (2010)
11. J. Ching, A. Wong, K. Chan, Class-dependent discretization for inductive learning from continuous and mixed-mode data. IEEE Trans. Pattern Anal. Mach. Intell. **17**(7), 641–651 (1995)
12. P. Clark, T. Niblett, The CN2 induction algorithm. Mach. Learn. **3**(4), 261–283 (1989)
13. B. Delibasic et al., Component-based decision trees for classification. Intell. Data Anal. **15**(5), 1–38 (2011)
14. F. Esposito, D. Malerba, G. Semeraro, A comparative analysis of methods for pruning decision trees. IEEE Trans. Pattern Anal. Mach. Intell. **19**(5), 476–491 (1997)
15. U. Fayyad, K. Irani, The attribute selection problem in decision tree generation, in *National Conference on Artificial Intelligence*. pp. 104–110 (1992)
16. A. Frank, A. Asuncion, UCI Machine Learning Repository (2010)
17. J.H. Friedman, A recursive partitioning decision rule for nonparametric classification. IEEE Trans. Comput. **100**(4), 404–408 (1977)
18. M. Gleser, M. Collen, Towards automated medical decisions. Comput. Biomed. Res. **5**(2), 180–189 (1972)
19. T. Ho, M. Basu, Complexity measures of supervised classification problems. IEEE Trans. Pattern Anal. Mach. Intell. **24**(3), 289–300 (2002)
20. T. Ho, M. Basu, M. Law, *Measures of Geometrical Complexity in Classification Problems*, Data Complexity in Pattern Recognition (Springer, London, 2006)
21. B. Jun et al., A new criterion in selection and discretization of attributes for the generation of decision trees. IEEE Trans. Pattern Anal. Mach. Intell. **19**(2), 1371–1375 (1997)
22. I. Kononenko, I. Bratko, E. Roskar, Experiments in automatic learning of medical diagnostic rules. Tech. rep. Ljubljana, Yugoslavia: Jozef Stefan Institute (1984)
23. W. Loh, Y. Shih, Split selection methods for classification trees. Stat. Sinica **7**, 815–840 (1997)
24. R.L. De Mántaras, *A Distance-Based Attribute Selection Measure for Decision Tree Induction*, Machine learning 6.1 (Kluwer, The Netherland, 1991). ISSN: 0885–6125
25. J. Martin, An exact probability metric for decision tree splitting and stopping. Mach. Learn. **28**(2), 257–291 (1997)
26. J. Mingers, Expert systems—rule induction with statistical data. J. Oper. Res. Soc. **38**, 39–47 (1987)

27. J. Mingers, An empirical comparison of selection measures for decision-tree induction. Mach. Learn. **3**(4), 319–342 (1989)
28. T. Niblett, I. Bratko, Learning decision rules in noisy domains, in *6th Annual Technical Conference on Research and Development in Expert Systems III*. pp. 25–34 (1986)
29. G.L. Pappa, Automatically Evolving Rule Induction Algorithms with Grammar-Based Genetic Programming. PhD thesis. University of Kent at Canterbury (2007)
30. G.L. Pappa, A.A. Freitas, *Automating the Design of Data Mining Algorithms: An Evolutionary Computation Approach* (Springer Publishing Company, Incorporated, 2009)
31. J.R. Quinlan, Induction of decision trees. Mach. Learn. **1**(1), 81–106 (1986)
32. J.R. Quinlan, Decision trees as probabilistic classifiers, in *4th International Workshop on Machine Learning* (1987)
33. J.R. Quinlan, Simplifying decision trees. Int. J. Man-Mach. Stud. **27**, 221–234 (1987)
34. J.R. Quinlan, Unknown attribute values in induction, in *6th International Workshop on Machine Learning*. pp. 164–168 (1989)
35. J. R. Quinlan, C4.5: programs for machine learning. San Francisco: Morgan Kaufmann (1993). ISBN: 1-55860-238-0
36. C.E. Shannon, A mathematical theory of communication. BELL Syst. Tech. J. **27**(1), 379–423, 625–56 (1948)
37. P.C. Taylor, B.W. Silverman, Block diagrams and splitting criteria for classification trees. Stat. Comput. **3**, 147–161 (1993)
38. A. Vella, D. Corne, C. Murphy, Hyper-heuristic decision tree induction, in *World Congress on Nature and Biologically Inspired Computing*, pp. 409–414 (2010)
39. I.H. Witten, E. Frank, Data mining: practical machine learning tools and techniques with java implementations. Morgan Kaufmann. ISBN: 1558605525 (1999)

Chapter 5
HEAD-DT: Experimental Analysis

Abstract In this chapter, we present several empirical analyses that assess the performance of HEAD-DT in different scenarios. We divide these analyses into two sets of experiments, according to the meta-training strategy employed for automatically designing the decision-tree algorithms. As mentioned in Chap. 4, HEAD-DT can operate in two distinct frameworks: (i) evolving a decision-tree induction algorithm tailored to one specific data set (specific framework); or (ii) evolving a decision-tree induction algorithm from multiple data sets (general framework). The specific framework provides data from a single data set to HEAD-DT for both algorithm design (evolution) and performance assessment. The experiments conducted for this scenario (see Sect. 5.1) make use of public data sets that do not share a common application domain. In the general framework, distinct data sets are used for algorithm design and performance assessment. In this scenario (see Sect. 5.2), we conduct two types of experiments, namely the *homogeneous approach* and the *heterogeneous approach*. In the *homogeneous approach*, we analyse whether automatically designing a decision-tree algorithm for a particular domain provides good results. More specifically, the data sets that feed HEAD-DT during evolution, and also those employed for performance assessment, share a common application domain. In the *heterogeneous approach*, we investigate whether HEAD-DT is capable of generating an algorithm that performs well across a variety of different data sets, regardless of their particular characteristics or application domain. We also discuss about the theoretic and empirical time complexity of HEAD-DT in Sect. 5.3, and we make a brief discussion on the cost-effectiveness of automated algorithm design in Sect. 5.4. We present examples of algorithms which were automatically designed by HEAD-DT in Sect. 5.5. We conclude the experimental analysis by empirically verifying in Sect. 5.6 whether the genetic search is worthwhile.

Keywords Experimental analysis · Specific framework · General framework · Cost-effectiveness of automatically-designed algorithms

© The Author(s) 2015
R.C. Barros et al., *Automatic Design of Decision-Tree Induction Algorithms*,
SpringerBriefs in Computer Science, DOI 10.1007/978-3-319-14231-9_5

Table 5.1 Summary of the data sets used in the experiments

Data set	# Instances	# Attributes	# Numeric attributes	# Nominal attributes	% Missing	# Classes
Abalone	4,177	8	7	1	0.00	30
Anneal	898	38	6	32	0.00	6
Arrhythmia	452	279	206	73	0.32	16
Audiology	226	69	0	69	2.03	24
Bridges_version1	107	12	3	9	5.53	6
Car	1,728	6	0	6	0.00	4
Cylinder_bands	540	39	18	21	4.74	2
Glass	214	9	9	0	0.00	7
Hepatitis	155	19	6	13	5.67	2
Iris	150	4	4	0	0.00	3
kdd_synthetic	600	61	60	1	0.00	6
Segment	2,310	19	19	0	0.00	7
Semeion	1,593	265	265	0	0.00	2
Shuttle_landing	15	6	0	6	28.89	2
Sick	3,772	30	6	22	5.54	2
Tempdiag	120	7	1	6	0.00	2
Tep.fea	3,572	7	7	0	0.00	3
Vowel	990	13	10	3	0.00	11
Winequality_red	1,599	11	11	0	0.00	10
Winequality_white	4,898	11	11	0	0.00	10

5.1 Evolving Algorithms Tailored to One Specific Data Set

In this first set of experiments, we investigate the performance of HEAD-DT with regard to the specific framework. For that, we employed 20 public data sets (see Table 5.1) that were collected from the UCI machine-learning repository[1] [8]. We compare the resulting decision-tree algorithms with the two most well-known and widely-used decision-tree induction algorithms: C4.5 [11] and CART [4]. We report the classification accuracy of the decision trees generated for each data set, as well as the F-Measure, and the size of the decision tree (total number of nodes). All results are based on the average of 10-fold cross-validation runs. Additionally, since HEAD-DT is a non-deterministic method, we averaged the results of 5 different runs (varying the random seed).

The baseline algorithms are configured with default parameters. We did not perform a parameter optimisation procedure in this set of experiments, given that we are designing one algorithm per data set, and optimising a set of parameters for each

---

[1] http://archive.ics.uci.edu/ml/.

data set is not feasible. Thus, we employed typical values found in the literature of evolutionary algorithms for decision-tree induction:

- Fitness-function: F-Measure;
- Population size: 100;
- Maximum number of generations: 100;
- Selection: tournament selection with size $t = 2$;
- Elitism rate: 5 individuals;
- Crossover: uniform crossover with 90 % probability;
- Mutation: random uniform gene mutation with 5 % probability;
- Reproduction: cloning individuals with 5 % probability.

In order to provide some reassurance about the validity and non-randomness of the obtained results, we present the results of statistical tests by following the approach proposed by Demšar [7]. In brief, this approach seeks to compare multiple algorithms on multiple data sets, and it is based on the use of the Friedman test with a corresponding post-hoc test. The Friedman test is a non-parametric counterpart of the well-known ANOVA, as follows. Let R_i^j be the rank of the jth of k algorithms on the ith of N data sets. The Friedman test compares the average ranks of algorithms, $R_j = \frac{1}{N} \sum_i R_i^j$. The Friedman statistic is given by:

$$\chi_F^2 = \frac{12N}{k(k+1)} \left[\sum_j R_j^2 - \frac{k(k+1)^2}{4} \right],$$ (5.1)

and it is distributed according to χ_F^2 with $k - 1$ degrees of freedom, when N and k are big enough.

Iman and Davenport [9] showed that Friedman's χ_F^2 is undesirably conservative and derived an adjusted statistic:

$$F_f = \frac{(N-1) \times \chi_F^2}{N \times (k-1) - \chi_F^2}$$ (5.2)

which is distributed according to the F-distribution with $k - 1$ and $(k - 1)(N - 1)$ degrees of freedom.

If the null hypothesis of similar performances is rejected, then we proceed with the Nemenyi post-hoc test for pairwise comparisons. The performance of two classifiers is significantly different if their corresponding average ranks differ by at least the critical difference

$$CD = q_\alpha \sqrt{\frac{k(k+1)}{6N}}$$ (5.3)

where critical values q_α are based on the Studentized range statistic divided by $\sqrt{2}$.

Table 5.2 Classification accuracy of CART, C4.5 and HEAD-DT

	CART	C4.5	HEAD-DT
Abalone	**0.26 ± 0.02**	0.22 ± 0.02	0.20 ± 0.02
Anneal	0.98 ± 0.01	0.99 ± 0.01	0.99 ± 0.01
Arrhythmia	**0.71 ± 0.05**	0.65 ± 0.04	0.65 ± 0.04
Audiology	0.74 ± 0.05	0.78 ± 0.07	**0.80 ± 0.06**
bridges_version1	0.41 ± 0.07	0.57 ± 0.10	**0.60 ± 0.12**
Car	0.97 ± 0.02	0.93 ± 0.02	**0.98 ± 0.01**
Cylinder_bands	0.60 ± 0.05	0.58 ± 0.01	**0.72 ± 0.04**
Glass	0.70 ± 0.11	0.69 ± 0.04	**0.73 ± 0.10**
Hepatitis	0.79 ± 0.05	0.79 ± 0.06	**0.81 ± 0.08**
Iris	0.93 ± 0.05	0.94 ± 0.07	**0.95 ± 0.04**
kdd_synthetic	0.88 ± 0.00	0.91 ± 0.04	**0.97 ± 0.03**
Segment	0.96 ± 0.01	0.97 ± 0.01	0.97 ± 0.01
Semeion	0.94 ± 0.01	0.95 ± 0.02	**1.00 ± 0.00**
Shuttle_landing	0.95 ± 0.16	0.95 ± 0.16	**0.95 ± 0.15**
Sick	0.99 ± 0.01	0.99 ± 0.00	0.99 ± 0.00
Tempdiag	1.00 ± 0.00	1.00 ± 0.00	1.00 ± 0.00
Tep.fea	0.65 ± 0.02	0.65 ± 0.02	0.65 ± 0.02
Vowel	0.82 ± 0.04	0.83 ± 0.03	**0.89 ± 0.04**
Winequality_red	0.63 ± 0.02	0.61 ± 0.03	**0.64 ± 0.03**
Winequality_white	0.58 ± 0.02	0.61 ± 0.03	**0.63 ± 0.03**

Table 5.2 shows the classification accuracy of C4.5, CART, and HEAD-DT. It illustrates the average accuracy over the 10-fold cross-validation runs ± the standard deviation of the accuracy obtained in those runs (best absolute values in bold). It is possible to see that HEAD-DT generates more accurate trees in 13 out of the 20 data sets. CART provides more accurate trees in two data sets, and C4.5 in none. In the remaining 5 data sets, no method was superior to the others.

To evaluate the statistical significance of the accuracy results, we calculated the average rank for CART, C4.5 and HEAD-DT: 2.375, 2.2 and 1.425, respectively. The average rank suggests that HEAD-DT is the best performing method regarding accuracy. The calculation of Friedman's χ_F^2 is given by:

$$\chi_F^2 = \frac{12 \times 20}{3 \times 4} \left[2.375^2 + 2.2^2 + 1.425^2 - \frac{3 \times 4^2}{4} \right] = 10.225 \qquad (5.4)$$

Iman's F statistic is given by:

$$F_f = \frac{(20 - 1) \times 10.225}{20 \times (3 - 1) - 10.225} = 6.52 \qquad (5.5)$$

Critical value of $F(k-1, (k-1)(n-1)) = F(2, 38)$ for $\alpha = 0.05$ is 3.25. Since $F_f > F_{0.05}(2, 38)$ (6.52 > 3.25), the null-hypothesis is rejected. We proceed with a post-hoc Nemenyi test to find which method provides better results. The critical difference CD is given by:

$$CD = 2.343 \times \sqrt{\frac{3 \times 4}{6 \times 20}} = 0.74 \tag{5.6}$$

The difference between the average rank of HEAD-DT and C4.5 is 0.775, and between HEAD-DT and CART is 0.95. Since both the differences are greater than CD, the performance of HEAD-DT is significantly better than both C4.5 and CART regarding accuracy.

Table 5.3 shows the classification F-Measure of C4.5, CART and HEAD-DT. The experimental results show that HEAD-DT generates better trees (regardless of the class imbalance problem) in 16 out of the 20 data sets. CART generates the best tree in two data sets, while C4.5 does not provide the best tree for any data set.

We calculated the average rank for CART, C4.5 and HEAD-DT: 2.5, 2.225 and 1.275, respectively. The average rank suggest that HEAD-DT is the best performing method regarding F-Measure. The calculation of Friedman's χ_F^2 is given by:

Table 5.3 Classification F-Measure of CART, C4.5 and HEAD-DT

	CART	C4.5	HEAD-DT
Abalone	**0.22 ± 0.02**	0.21 ± 0.02	0.20 ± 0.02
Anneal	0.98 ± 0.01	0.98 ± 0.01	**0.99 ± 0.01**
Arrhythmia	**0.67 ± 0.05**	0.64 ± 0.05	0.63 ± 0.06
Audiology	0.70 ± 0.04	0.75 ± 0.08	**0.79 ± 0.07**
Bridges_version1	0.44 ± 0.06	0.52 ± 0.10	**0.56 ± 0.12**
Car	0.93 ± 0.97	0.93 ± 0.02	**0.98 ± 0.01**
Cylinder_bands	0.54 ± 0.07	0.42 ± 0.00	**0.72 ± 0.04**
Glass	0.67 ± 0.10	0.67 ± 0.05	**0.72 ± 0.09**
Hepatitis	0.74 ± 0.07	0.77 ± 0.06	**0.80 ± 0.08**
Iris	0.93 ± 0.05	0.93 ± 0.06	**0.95 ± 0.05**
kdd_synthetic	0.88 ± 0.03	0.90 ± 0.04	**0.97 ± 0.03**
Segment	0.95 ± 0.01	0.96 ± 0.09	**0.97 ± 0.01**
Semeion	0.93 ± 0.01	0.95 ± 0.02	**1.00 ± 0.00**
Shuttle_landing	0.56 ± 0.03	0.56 ± 0.38	**0.93 ± 0.20**
Sick	0.98 ± 0.00	0.98 ± 0.00	**0.99 ± 0.00**
Tempdiag	1.00 ± 0.00	1.00 ± 0.00	1.00 ± 0.00
Tep.fea	0.60 ± 0.02	0.61 ± 0.02	0.61 ± 0.02
Vowel	0.81 ± 0.03	0.82 ± 0.03	**0.89 ± 0.03**
Winequality_red	0.61 ± 0.02	0.60 ± 0.03	**0.63 ± 0.03**
Winequality_white	0.57 ± 0.02	0.60 ± 0.02	**0.63 ± 0.03**

$$\chi_F^2 = \frac{12 \times 20}{3 \times 4}\left[2.5^2 + 2.225^2 + 1.275^2 - \frac{3 \times 4^2}{4}\right] = 16.525 \quad (5.7)$$

Iman's F statistic is given by:

$$F_f = \frac{(20 - 1) \times 16.525}{20 \times (3 - 1) - 16.525} = 13.375 \quad (5.8)$$

Since $F_f > F_{0.05}(2, 38)$ (13.375 > 3.25), the null-hypothesis is rejected. The difference between the average rank of HEAD-DT and C4.5 is 0.95 and that between HEAD-DT and CART is 1.225. Since both the differences are greater than CD (0.74), the performance of HEAD-DT is significantly better than both C4.5 and CART regarding F-Measure.

Table 5.4 shows the size of trees generated by C4.5, CART and HEAD-DT. Results show that CART generates smaller trees in 15 out of the 20 data sets. C4.5 generates smaller trees in 2 data sets, and HEAD-DT in only one data set. The statistical analysis is given as follows:

Table 5.4 Tree size of CART, C4.5 and HEAD-DT trees

	CART	C4.5	HEAD-DT
Abalone	**44.40 ± 16.00**	2088.90 ± 37.63	4068.12 ± 13.90
Anneal	**21.00 ± 3.13**	48.30 ± 6.48	55.72 ± 3.66
Arrhythmia	**23.20 ± 2.90**	82.60 ± 5.80	171.84 ± 5.18
Audiology	**35.80 ± 11.75**	50.40 ± 4.01	118.60 ± 3.81
Bridges_version1	**1.00 ± 0.00**	24.90 ± 20.72	156.88 ± 14.34
Car	**108.20 ± 16.09**	173.10 ± 6.51	171.92 ± 4.45
Cylinder_bands	4.20 ± 1.03	**1.00 ± 0.00**	211.44 ± 9.39
Glass	**23.20 ± 10.56**	44.80 ± 5.20	86.44 ± 3.14
Hepatitis	**6.60 ± 8.58**	15.40 ± 4.40	71.80 ± 4.77
Iris	**6.20 ± 1.69**	8.00 ± 1.41	20.36 ± 1.81
kdd_synthetic	**1.00 ± 0.00**	37.80 ± 4.34	26.16 ± 2.45
Segment	**78.00 ± 8.18**	80.60 ± 4.97	132.76 ± 3.48
Semeion	34.00 ± 12.30	55.00 ± 8.27	**19.00 ± 0.00**
Shuttle_landing	1.00 ± 0.00	1.00 ± 0.00	5.64 ± 1.69
Sick	**45.20 ± 11.33**	46.90 ± 9.41	153.70 ± 8.89
Tempdiag	5.00 ± 0.00	5.00 ± 0.00	5.32 ± 1.04
Tep.fea	13.00 ± 2.83	**8.20 ± 1.69**	18.84 ± 1.97
Vowel	**175.80 ± 23.72**	220.70 ± 20.73	361.42 ± 5.54
Winequality_red	**151.80 ± 54.58**	387.00 ± 26.55	796.00 ± 11.22
Winequality_white	**843.80 ± 309.01**	1367.20 ± 58.44	2525.88 ± 13.17

$$\chi_F^2 = \frac{12 \times 20}{3 \times 4} \left[1.2^2 + 2^2 + 2.8^2 - \frac{3 \times 4^2}{4} \right] = 25.6 \qquad (5.9)$$

$$F_f = \frac{(20 - 1) \times 33.78}{20 \times (3 - 1) - 25.6} = 33.78 \qquad (5.10)$$

Since $F_f > F_{0.05}(2, 38)$ (33.78 > 3.25), the null-hypothesis is rejected. The difference between the average rank of HEAD-DT and C4.5 is 0.8 and that between HEAD-DT and CART is 1.6. Since both the differences are greater than CD (0.74), HEAD-DT generates trees which are significantly larger than both C4.5 and CART. This should not be a concern, since smaller trees are only preferable in scenarios where the predictive performance of the algorithms is similar.

The statistical analysis previously presented clearly indicates that HEAD-DT generates algorithms whose trees outperform C4.5 and CART regarding predictive performance. The Occam's razor assumption that among competitive hypotheses, the simpler is preferred, does not apply to this case.

5.2 Evolving Algorithms from Multiple Data Sets

In this section, we evaluate the performance of HEAD-DT when evolving a single algorithm from multiple data sets (the general framework). More specifically, we divide this experiment into two scenarios: (i) evolving a single decision-tree algorithm for data sets from a particular application domain (homogeneous approach); and (ii) evolving a single decision-tree algorithm for a variety of data sets (heterogeneous approach).

In both cases, we need to establish the methodology to select the data sets that will compose the meta-training and meta-test sets. Hence, we developed the following selection methodology: randomly choose 1 data set from the available set to be part of the meta-training set; then, execute HEAD-DT with the selected data set in the meta-training set and the remaining data sets in the meta-test set; for the next experiment, select two additional data sets that were previously part of the meta-test set, and move them to the meta-training set, that now comprises 3 data sets. This procedure is repeated until we have 9 data sets being part of the meta-training set.

Considering that HEAD-DT is a regular generational EA (as depicted in Fig. 4.1), the following parameters have to be chosen prior to evolution: (i) population size; (ii) maximum number of generations; (iii) tournament selection size; (iv) elitism rate; (v) reproduction probability; (vi) crossover probability; and (vii) mutation probability.

For parameters (i)–(iv), we defined values commonly used in the literature of evolutionary algorithms for decision-tree induction [2], namely: 100 individuals, 100 generations, tournament between 2 individuals, 5 % of elitism. For the remaining parameters, since the selected individuals will undergo either reproduction, crossover or mutation (mutually exclusive operators), we employ the single parameter p that was previously presented in Chap. 4, Sect. 4.3.

For both homogeneous and heterogeneous approaches, we performed a tuning procedure varying p within $\{0.1, 0.2, 0.3, 0.4, 0.5, 0.6, 0.7, 0.8, 0.9\}$. In this tuning procedure, we employed a particular set of data sets to be part of the *parameter optimisation* group in order to evaluate the "optimal" value of p. Note that the aim of the parameter optimisation procedure is not to optimise the parameters for a particular data set, but to find robust values that work well across the tuning data sets. We then use the robust value of p found in the procedure in another set of data sets that were selected to be part of the experiments. This evaluates the generalisation ability of p across new data sets, unused for parameter tuning, as usual in supervised machine learning.

Also for both approaches, the fitness function employed by HEAD-DT is the average F-Measure of the data sets belonging to the meta-training set. To evaluate the performance of the best algorithm evolved by HEAD-DT, we performed a 10-fold cross-validation procedure for each data set belonging to the meta-test set, recording the accuracy and F-Measure achieved by each of the corresponding decision trees. Additionally, to mitigate the randomness effect of evolutionary algorithms, we average the results of 5 different runs of HEAD-DT.

In this framework, we once again make use of CART and C4.5 as the baseline algorithms in the experimental analysis. We employ their java versions available from the Weka machine learning toolkit [14] under the names of SimpleCART and J48. Moreover, we also compare HEAD-DT with the REPTree algorithm, which is a variation of C4.5 that employs reduced-error pruning, also available from the Weka toolkit.

For all baseline algorithms, we employ their default parameters, since they were carefully optimised by their respective authors throughout several years. Note that, in the supervised machine learning literature, the common approach is to find the optimal parameters for being used in a variety of distinct data sets, instead of optimising the algorithm for each specific data set.

5.2.1 The Homogeneous Approach

In this set of experiments, we assess the relative performance of the algorithm automatically designed by HEAD-DT for data sets from a particular application domain. More specifically, we make use of 35 publicly-available data sets from microarray gene expression data, described in [12]. In brief, microarray technology enables expression level measurement for thousands of genes in parallel, given a biological tissue of interest. Once combined, results from a set of microarray experiments produce a gene expression data set. The data sets employed here are related to different types or subtypes of cancer, e.g., patients with prostate, lung, skin, and other types of cancer. The classification task refers to labeling different examples (instances) according to their gene (attribute) expression levels. The main structural characteristics of the 35 datasets are summarized in Table 5.5.

Table 5.5 Summary of the 35 Gene Expression data sets

	Data set	Chip	# Instances	# Attributes	IR	# Classes
Parameter optimisation	Armstrong-v2	Affy	72	2193	1.40	3
	Bredel	cDNA	50	1,738	6.20	3
	Dyrskjot	Affy	40	1,202	2.22	3
	Garber	cDNA	66	4,552	10.00	4
	Golub-v2	Affy	72	1,867	4.22	3
	Gordon	Affy	181	1,625	4.84	2
	Khan	cDNA	83	1,068	2.64	4
	Laiho	Affy	37	2,201	3.63	2
	Pomeroy-v2	Affy	42	1,378	2.50	5
	Ramaswamy	Affy	190	1,362	3.00	14
	Su	Affy	174	1,570	4.67	10
	Tomlins-v2	cDNA	92	1,287	2.46	4
	Yeoh-v1	Affy	248	2,525	4.77	2
	Yeoh-v2	Affy	248	2,525	5.27	6
Experiments	Alizadeh-v1	cDNA	42	1,094	1.00	2
	Alizadeh-v2	cDNA	62	2,092	4.67	3
	Alizadeh-v3	cDNA	62	2,092	2.33	4
	Armstrong-v1	Affy	72	1,080	2.00	2
	Bhattacharjee	Affy	203	1,542	23.17	5
	Bittner	cDNA	38	2,200	1.00	2
	Chen	cDNA	179	84	1.39	2
	Chowdary	Affy	104	181	1.48	2
	Golub-v1	Affy	72	1,867	1.88	2
	Lapointe-v1	cDNA	69	1,624	3.55	3
	Lapointe-v2	cDNA	110	2,495	3.73	4
	Liang	cDNA	37	1,410	9.34	3
	Nutt-v1	Affy	50	1,376	2.14	4
	Nutt-v2	Affy	28	1,069	1.00	2
	Nutt-v3	Affy	22	1,151	2.14	2
	Shipp-v1	Affy	77	797	3.05	2
	Pomeroy-v1	Affy	34	856	2.78	2
	Risinger	cDNA	42	1,770	6.33	4
	Singh	Affy	102	338	1.04	2
	Tomlins-v1	cDNA	104	2,314	2.67	5
	West	Affy	49	1,197	1.04	2

For each data set, we present type of microarray chip, the total number of instances, total number of attributes, imbalanced ratio (rate between over- and under-represented class), and total number of classes

It is important to point out that microarray technology is generally available in two different types of platforms: single-channel microarrays (e.g., Affymetrix) or double-channel microarrays (e.g., cDNA). The type of microarray chip from each data set is in the second column of Table 5.5. Measurements of Affymetrix arrays are estimates on the number of RNA copies found in the cell sample, whereas cDNA microarrays values are ratios of the number of copies in relation to a control cell sample. As in [12], all genes with expression level below 10 are set to a minimum threshold of 10 in the Affymetrix data. The maximum threshold is set to 16,000. This is because values below or above these thresholds are often said to be unreliable [10]. Thus, the experimental analysis is performed on the scaled data to which the ceiling and threshold values have been applied. Still for the case of Affymetrix data, the following procedure is applied in order to remove uninformative genes: for each gene j (attribute), compute the mean m_j among the samples (instances). In order to get rid of extreme values, the first 10 % largest and smallest values are discarded. Based on such a mean, every value x_{ij} of gene i and sample j is transformed as follows: $y_{ij} = \log_2(x_{ij}/m_j)$. We then selected genes with expression levels differing by at least l-fold in at least c samples from their mean expression level across the samples. With few exceptions, the parameters l and c were selected in order to produce a filtered data set with at least 10 % of the original number of genes.[2] It should be noticed that the transformed data is only used in the filtering step. A similar filtering procedure was applied for the cDNA data, but without the log transformation. In the case of cDNA microarray data sets, genes with more than 10 % of missing values were discarded. The remaining genes that still presented missing values had them replaced by its respective mean value.

Note that we randomly divided the 35 data sets into two groups: *parameter optimisation* and *experiments*. The 14 data sets in the *parameter optimisation* group are used for tuning the evolutionary parameters of HEAD-DT. The remaining 21 data sets from the *experiments* group are used for evaluating the performance of the algorithms automatically designed by HEAD-DT.

Following the selection methodology previously presented, the 14 *parameter optimisation* data sets are arranged in 5 different experimental configurations {#training sets, #test sets}: $\{1 \times 13\}$, $\{3 \times 11\}$, $\{5 \times 9\}$, $\{7 \times 7\}$, and $\{9 \times 5\}$. Similarly, the 21 data sets in the *experiments* group are arranged in also 5 different experimental configurations {#training sets, #test sets}: $\{1 \times 20\}$, $\{3 \times 18\}$, $\{5 \times 16\}$, $\{7 \times 14\}$, and $\{9 \times 12\}$. Table 5.6 presents the randomly selected data sets according to the configurations detailed.

5.2.1.1 Parameter Optimisation

Table 5.7 presents the results of the tuning experiments. We present the average ranking of each version of HEAD-DT (H-p) in the corresponding experimental

---

[2] The values of l and c for each data set can be found at http://algorithmics.molgen.mpg.de/Static/Supplements/CompCancer/datasets.htm.

Table 5.6 Meta-training and meta-test configurations for the gene expression data

Data sets for parameter optimisation

{1 × 13}		{3 × 11}		{5 × 9}		{7 × 7}		{9 × 5}	
Meta-training	Meta-test	Meta-training	Meta-test	Meta-training	Meta-test	Meta-training	Meta-test	Meta-training	Meta-test
Bredel	Armstrong-v2	Bredel	Armstrong-v2	Bredel	Armstrong-v2	Bredel	Dyrskjot	Bredel	Dyrskjot
	Dyrskjot	Pomeroy-v2	Dyrskjot	Pomeroy-v2	Dyrskjot	Pomeroy-v2	Garber	Pomeroy-v2	Garber
	Garber	Yeoh-v2	Garber	Yeoh-v2	Garber	Yeoh-v2	Golub-v2	Yeoh-v2	Golub-v2
	Golub-v2		Golub-v2	Gordon	Golub-v2	Gordon	Khan	Gordon	Ramaswamy
	Gordon		Gordon	Su	Khan	Su	Ramaswamy	Su	Yeoh-v1
	Khan		Khan		Laiho	Armstrong-v2	Tomlins-v2	Armstrong-v2	
	Laiho		Laiho		Ramaswamy	Laiho	Yeoh-v1	Laiho	
	Pomeroy-v2		Ramaswamy		Su			Khan	
	Ramaswamy		Su		Tomlins-v2			Tomlins-v2	
	Su		Tomlins-v2		Yeoh-v1				
	Tomlins-v2		Yeoh-v1						
	Yeoh-v1								
	Yeoh-v2								

Data sets for the experimental analysis

{1 × 20}		{3 × 18}		{5 × 16}		{7 × 14}		{9 × 12}	
Meta-training	Meta-test	Meta-training	Meta-test	Meta-training	Meta-test	Meta-training	Meta-test	Meta-training	Meta-test
Nutt-v3	Alizadeh-v1		Alizadeh-v1		Alizadeh-v1	Nutt-v3	Alizadeh-v1	Nutt-v3	Alizadeh-v1
	Alizadeh-v2		Alizadeh-v2		Alizadeh-v2	Alizadeh-v3	Alizadeh-v2	Alizadeh-v3	Alizadeh-v2
	Alizadeh-v3		Armstrong-v1		Bhattacharjee		Bhattacharjee	Lapointe-v2	Bhattacharjee
	Armstrong-v1		Bhattacharjee		Bittner		Bittner	Chowdary	Chen
	Bhattacharjee		Bittner		Chen		Chen		Golub-v1

(continued)

Table 5.6 (continued)

Data sets for the experimental analysis

{1 × 20}		{3 × 18}		{5 × 16}		{7 × 14}		{9 × 12}	
Meta-training	Meta-test	Meta-training	Meta-test	Meta-training	Meta-test	Meta-training	Meta-test	Meta-training	Meta-test
	Bittner	Nutt-v3	Chen	Nutt-v3	Chowdary	Lapointe-v2	Golub-v1	Armstrong-v1	Lapointe-v1
	Chen	Alizadeh-v3	Chowdary	Alizadeh-v3	Golub-v1	Armstrong-v1	Lapointe-v1	Tomlins-v1	Liang
	Chowdary	Lapointe-v2	Golub-v1	Lapointe-v2	Lapointe-v1	Tomlins-v1	Liang	Bittner	Nutt-v1
	Golub-v1		Lapointe-v1	Armstrong-v1	Liang	Bittner	Nutt-v1	Risinger	Nutt-v2
	Lapointe-v1		Liang	Tomlins-v1	Nutt-v1	Risinger	Nutt-v2	Chowdary	Shipp-v1
	Lapointe-v2		Nutt-v1		Nutt-v2		Shipp-v1	West	Pomeroy-v1
	Liang		Nutt-v2		Shipp-v1		Pomeroy-v1		Singh
	Nutt-v1		Shipp-v1		Pomeroy-v1		Singh		
	Nutt-v2		Pomeroy-v1		Risinger		West		
	Shipp-v1		Risinger		Singh				
	Pomeroy-v1		Singh		West				
	Risinger		Tomlins-v1						
	Singh		West						
	Tomlins-v1								
	West								

Data sets were randomly selected according to the methodology presented in the beginning of Sect. 5.2

Table 5.7 Results of the tuning experiments for the homogeneous

Configuration	Rank	H-0.1	H-0.2	H-0.3	H-0.4	H-0.5	H-0.6	H-0.7	H-0.8	H-0.9
{1 × 13}	Accuracy	5.73	4.23	4.54	4.92	5.19	4.42	5.31	5.62	5.04
	F-measure	6.15	4.85	4.31	5.00	5.12	4.12	4.75	5.62	5.08
{3 × 11}	Accuracy	4.91	4.41	4.59	4.45	4.68	5.27	5.64	5.64	5.41
	F-measure	4.68	4.45	4.09	4.59	5.05	5.95	5.27	6.14	4.77
{5 × 9}	Accuracy	5.11	6.17	5.22	6.28	4.28	5.11	4.61	3.44	4.77
	F-measure	4.72	6.06	4.72	6.06	4.00	5.67	4.89	3.89	5.00
{7 × 7}	Accuracy	6.07	7.43	4.71	6.50	4.86	3.36	3.79	3.50	4.79
	F-measure	6.21	6.71	5.14	3.21	5.14	3.71	4.00	3.21	4.50
{9 × 5}	Accuracy	8.30	5.70	5.50	5.20	7.90	2.70	4.00	2.80	2.9
	F-measure	7.10	5.60	5.60	4.90	6.70	3.50	4.50	4.20	2.90
	Average	5.90	5.56	4.84	5.11	5.29	**4.38**	4.68	4.41	4.52

HEAD-DT is executed with different values of parameter p(H-p). Values are the average performance (rank) of each HEAD-DT version in the corresponding meta-test set of tuning data sets, according to either accuracy or F-Measure

configuration. For example, when H-0.1 makes use of a single data set for evolving the optimal algorithm ({1 × 13}), its performance in the remaining 13 data sets gives H-0.1 the average rank position of 5.73 regarding the accuracy of its corresponding decision trees, and 6.15 regarding F-Measure.

The Friedman and Nemenyi tests did not indicate any statistically significant differences among the 9 distinct versions of HEAD-DT, either considering accuracy or F-Measure, for any of the experimental configurations. This lack of significant differences indicates that HEAD-DT is robust to different values of p. For selecting the ideal value of p to employ in the experimental analysis, we averaged the results across the different configurations, and also between the two different evaluation measures, accuracy and F-Measure. Hence, we calculated the average of the average ranks for each H-p version across the distinct configurations and evaluation measures, and the results are presented in the bottom of Table 5.7. It is possible to see how marginal are the differences among values of $p \geq 0.6$. H-0.6 was then selected as the optimised version of HEAD-DT, bearing in mind it presented the lowest average of the average ranks (4.38).

In the next section, we present the results for the experimental analysis performed over the 21 data sets in the *experiments* group, in which we compare H-0.6 (hereafter called simply HEAD-DT) to the baseline algorithms.

5.2.1.2 Experimental Results

Tables 5.8, 5.9, 5.10, 5.11 and 5.12 show the average values of accuracy and F-Measure achieved by HEAD-DT, CART, C4.5, and REPTree in configurations {1 × 20}, {3 × 18}, {5 × 16}, {7 × 14}, and {9 × 12}. At the bottom of each table,

Table 5.8 Results for the $\{1 \times 20\}$ configuration

(a) *Accuracy results*

Data set	HEAD-DT	CART	C4.5	REP
Alizadeh-2000-v1	0.77 ± 0.11	0.71 ± 0.16	0.68 ± 0.20	0.72 ± 0.19
Alizadeh-2000-v2	0.88 ± 0.06	0.92 ± 0.11	0.86 ± 0.15	0.84 ± 0.17
Alizadeh-2000-v3	0.71 ± 0.05	0.69 ± 0.15	0.71 ± 0.15	0.66 ± 0.13
Armstrong-2002-v1	0.90 ± 0.04	0.90 ± 0.07	0.89 ± 0.06	0.91 ± 0.07
Bhattacharjee-2001	0.90 ± 0.03	0.89 ± 0.10	0.89 ± 0.08	0.86 ± 0.06
Bittner-2000	0.61 ± 0.09	0.53 ± 0.18	0.49 ± 0.16	0.71 ± 0.22
Chen-2002	0.85 ± 0.04	0.85 ± 0.07	0.81 ± 0.07	0.79 ± 0.12
Chowdary-2006	0.95 ± 0.03	0.97 ± 0.05	0.95 ± 0.05	0.94 ± 0.06
Golub-1999-v1	0.88 ± 0.03	0.86 ± 0.07	0.88 ± 0.08	0.90 ± 0.10
Lapointe-2004-v1	0.66 ± 0.08	0.78 ± 0.18	0.71 ± 0.14	0.63 ± 0.09
Lapointe-2004-v2	0.62 ± 0.05	0.68 ± 0.19	0.57 ± 0.11	0.61 ± 0.15
Liang-2005	0.89 ± 0.09	0.71 ± 0.14	0.76 ± 0.19	0.78 ± 0.18
Nutt-2003-v1	0.53 ± 0.08	0.54 ± 0.19	0.50 ± 0.17	0.40 ± 0.21
Nutt-2003-v2	0.84 ± 0.08	0.77 ± 0.21	0.87 ± 0.17	0.45 ± 0.27
Pomeroy-2002-v1	0.88 ± 0.08	0.84 ± 0.17	0.88 ± 0.16	0.73 ± 0.10
Risinger-2003	0.58 ± 0.15	0.52 ± 0.15	0.53 ± 0.19	0.59 ± 0.21
Shipp-2002-v1	0.91 ± 0.05	0.77 ± 0.12	0.84 ± 0.16	0.77 ± 0.05
Singh-2002	0.78 ± 0.04	0.74 ± 0.14	0.78 ± 0.12	0.69 ± 0.15
Tomlins-2006	0.59 ± 0.07	0.58 ± 0.20	0.58 ± 0.18	0.58 ± 0.10
West-2001	0.89 ± 0.08	0.89 ± 0.11	0.84 ± 0.13	0.74 ± 0.19
Average rank	**1.75**	2.45	2.70	3.10

(b) *F-Measure results*

Data set	HEAD-DT	CART	C4.5	REP
Alizadeh-2000-v1	0.77 ± 0.11	0.67 ± 0.21	0.63 ± 0.26	0.72 ± 0.19
Alizadeh-2000-v2	0.89 ± 0.06	0.92 ± 0.10	0.84 ± 0.17	0.80 ± 0.19
Alizadeh-2000-v3	0.71 ± 0.05	0.65 ± 0.18	0.68 ± 0.17	0.63 ± 0.12
Armstrong-2002-v1	0.90 ± 0.04	0.90 ± 0.07	0.88 ± 0.06	0.91 ± 0.08
Bhattacharjee-2001	0.89 ± 0.02	0.87 ± 0.12	0.89 ± 0.08	0.85 ± 0.07
Bittner-2000	0.61 ± 0.09	0.49 ± 0.21	0.45 ± 0.19	0.66 ± 0.28
Chen-2002	0.85 ± 0.04	0.85 ± 0.07	0.81 ± 0.07	0.78 ± 0.13
Chowdary-2006	0.95 ± 0.03	0.97 ± 0.05	0.95 ± 0.05	0.94 ± 0.07
Golub-1999-v1	0.88 ± 0.03	0.86 ± 0.07	0.87 ± 0.08	0.90 ± 0.10
Lapointe-2004-v1	0.64 ± 0.10	0.73 ± 0.20	0.68 ± 0.15	0.50 ± 0.13
Lapointe-2004-v2	0.62 ± 0.05	0.66 ± 0.20	0.58 ± 0.10	0.57 ± 0.15
Liang-2005	0.87 ± 0.10	0.67 ± 0.21	0.77 ± 0.22	0.71 ± 0.23
Nutt-2003-v1	0.53 ± 0.08	0.48 ± 0.19	0.46 ± 0.16	0.31 ± 0.22
Nutt-2003-v2	0.84 ± 0.08	0.73 ± 0.26	0.87 ± 0.17	0.34 ± 0.31
Pomeroy-2002-v1	0.87 ± 0.10	0.79 ± 0.22	0.85 ± 0.20	0.62 ± 0.14

(continued)

Table 5.8 (continued)

(b) *F-Measure results*

Data set	HEAD-DT	CART	C4.5	REP
Risinger-2003	0.56 ± 0.15	0.48 ± 0.19	0.53 ± 0.21	0.52 ± 0.25
Shipp-2002-v1	0.90 ± 0.05	0.75 ± 0.13	0.83 ± 0.17	0.70 ± 0.10
Singh-2002	0.78 ± 0.04	0.72 ± 0.18	0.77 ± 0.13	0.67 ± 0.18
Tomlins-2006	0.58 ± 0.08	0.56 ± 0.20	0.56 ± 0.20	0.53 ± 0.11
West-2001	0.89 ± 0.08	0.89 ± 0.12	0.81 ± 0.17	0.71 ± 0.23
Average rank	**1.55**	2.50	2.60	3.35

we present the average rank position (the average of the rank position in each data set) of each method. The lower the ranking, the better the method. A method capable of outperforming any other method in every data set would have an average rank position of 1.00 (first place).

The first experiment was performed for configuration $\{1 \times 20\}$. Table 5.8 shows the results for this configuration considering accuracy (Table 5.8a) and F-Measure (Table 5.8b). Note that HEAD-DT is the best performing method with respect to both accuracy and F-Measure, reaching an average rank of 1.75 and 1.55, respectively. HEAD-DT is followed by CART (2.45 and 2.50) and C4.5 (2.70 and 2.65), whose performances are quite evenly matched. REPTree is the worst-ranked method for both accuracy (3.10) and F-Measure (3.35).

The next experiment concerns configuration $\{3 \times 18\}$, whose results are presented in Table 5.9. In this experiment, HEAD-DT's average rank is again the lowest of the experiment: 1.61 for both accuracy and F-Measure. That means HEAD-DT is often the best performing method (1st place) in the group of 18 test data sets. CART and C4.5 once again present very similar average rank values, which is not surprising bearing in mind they are both considered the state-of-the-art top-down decision-tree induction algorithms. REPTree is again the worst-performing method among the four algorithms.

Table 5.10 presents the results for configuration $\{5 \times 16\}$. The scenario is quite similar to the previous configurations, with HEAD-DT leading the ranking, with average rank values of 1.44 (accuracy) and 1.31 (F-Measure). These are the lowest rank values obtained by a method in any experimental configuration conducted in this analysis. HEAD-DT is followed by CART (2.69 and 2.69) and C4.5 (2.69 and 2.63), which once again present very similar performances. REPTree is at the bottom of the ranking, with 3.19 (accuracy) and 3.38 (F-Measure).

The experimental results for configuration $\{7 \times 14\}$ show a different picture. Table 5.11a, which depicts the accuracy values of each method, indicates that HEAD-DT is outperformed by both CART and C4.5, though their average rank values are very similar: 2.14, 2.21, versus 2.36 for HEAD-DT. REPTree keeps its position as worst-performing method, with an average rank value of 3.29. However, Table 5.11b returns to the same scenario presented in configurations $\{1 \times 20\}$, $\{3 \times 18\}$, and $\{5 \times 16\}$: HEAD-DT outperforming the baseline methods, with CART and C4.5 tied

Table 5.9 Results for the $\{3 \times 18\}$ configuration

(a) *Accuracy results*

Data set	HEAD-DT	CART	C4.5	REP
Alizadeh-2000-v1	0.74 ± 0.07	0.71 ± 0.16	0.68 ± 0.20	0.72 ± 0.19
Alizadeh-2000-v2	0.90 ± 0.07	0.92 ± 0.11	0.86 ± 0.15	0.84 ± 0.17
Armstrong-2002-v1	0.88 ± 0.01	0.90 ± 0.07	0.89 ± 0.06	0.91 ± 0.07
Bhattacharjee-2001	0.92 ± 0.03	0.89 ± 0.10	0.89 ± 0.08	0.86 ± 0.06
Bittner-2000	0.55 ± 0.09	0.53 ± 0.18	0.49 ± 0.16	0.71 ± 0.22
Chen-2002	0.87 ± 0.03	0.85 ± 0.07	0.81 ± 0.07	0.79 ± 0.12
Chowdary-2006	0.97 ± 0.01	0.97 ± 0.05	0.95 ± 0.05	0.94 ± 0.06
Golub-1999-v1	0.89 ± 0.02	0.86 ± 0.07	0.88 ± 0.08	0.90 ± 0.10
Lapointe-2004-v1	0.70 ± 0.06	0.78 ± 0.18	0.71 ± 0.14	0.63 ± 0.09
Liang-2005	0.89 ± 0.08	0.71 ± 0.14	0.76 ± 0.19	0.78 ± 0.18
Nutt-2003-v1	0.67 ± 0.10	0.54 ± 0.19	0.50 ± 0.17	0.40 ± 0.21
Nutt-2003-v2	0.77 ± 0.08	0.77 ± 0.21	0.87 ± 0.17	0.45 ± 0.27
Pomeroy-2002-v1	0.92 ± 0.05	0.84 ± 0.17	0.88 ± 0.16	0.73 ± 0.10
Risinger-2003	0.53 ± 0.11	0.52 ± 0.15	0.53 ± 0.19	0.59 ± 0.21
Shipp-2002-v1	0.85 ± 0.05	0.77 ± 0.12	0.84 ± 0.16	0.77 ± 0.05
Singh-2002	0.80 ± 0.04	0.74 ± 0.14	0.78 ± 0.12	0.69 ± 0.15
Tomlins-2006	0.64 ± 0.03	0.58 ± 0.20	0.58 ± 0.18	0.58 ± 0.10
West-2001	0.92 ± 0.04	0.90 ± 0.11	0.84 ± 0.13	0.74 ± 0.19
Average rank	**1.61**	2.61	2.72	3.05

(b) *F-Measure results*

Data set	HEAD-DT	CART	C4.5	REP
Alizadeh-2000-v1	0.73 ± 0.07	0.67 ± 0.21	0.63 ± 0.26	0.72 ± 0.19
Alizadeh-2000-v2	0.91 ± 0.06	0.92 ± 0.10	0.84 ± 0.17	0.80 ± 0.19
Armstrong-2002-v1	0.88 ± 0.02	0.90 ± 0.07	0.88 ± 0.06	0.91 ± 0.08
Bhattacharjee-2001	0.91 ± 0.03	0.87 ± 0.12	0.89 ± 0.08	0.85 ± 0.07
Bittner-2000	0.53 ± 0.10	0.49 ± 0.21	0.45 ± 0.19	0.66 ± 0.28
Chen-2002	0.87 ± 0.03	0.85 ± 0.07	0.81 ± 0.07	0.78 ± 0.13
Chowdary-2006	0.97 ± 0.01	0.97 ± 0.05	0.95 ± 0.05	0.94 ± 0.07
Golub-1999-v1	0.89 ± 0.02	0.86 ± 0.07	0.87 ± 0.08	0.90 ± 0.10
Lapointe-2004-v1	0.68 ± 0.08	0.73 ± 0.20	0.68 ± 0.15	0.50 ± 0.13
Liang-2005	0.87 ± 0.10	0.67 ± 0.21	0.77 ± 0.22	0.71 ± 0.23
Nutt-2003-v1	0.65 ± 0.10	0.48 ± 0.19	0.46 ± 0.16	0.31 ± 0.22
Nutt-2003-v2	0.77 ± 0.08	0.73 ± 0.26	0.87 ± 0.17	0.34 ± 0.31
Pomeroy-2002-v1	0.92 ± 0.05	0.79 ± 0.22	0.85 ± 0.20	0.62 ± 0.14
Risinger-2003	0.52 ± 0.11	0.48 ± 0.19	0.53 ± 0.21	0.52 ± 0.25
Shipp-2002-v1	0.84 ± 0.05	0.75 ± 0.13	0.83 ± 0.17	0.70 ± 0.10
Singh-2002	0.80 ± 0.04	0.72 ± 0.18	0.77 ± 0.13	0.67 ± 0.18
Tomlins-2006	0.63 ± 0.03	0.56 ± 0.20	0.56 ± 0.20	0.53 ± 0.11
West-2001	0.92 ± 0.04	0.89 ± 0.12	0.81 ± 0.17	0.71 ± 0.23
Average rank	**1.61**	2.61	2.56	3.22

Table 5.10 Results for the $\{5 \times 16\}$ configuration

(a) *Accuracy results*

Data set	HEAD-DT	CART	C4.5	REP
Alizadeh-2000-v1	0.75 ± 0.05	0.71 ± 0.16	0.68 ± 0.20	0.72 ± 0.19
Alizadeh-2000-v2	0.91 ± 0.07	0.92 ± 0.11	0.86 ± 0.15	0.84 ± 0.17
Bhattacharjee-2001	0.94 ± 0.02	0.89 ± 0.10	0.89 ± 0.08	0.86 ± 0.06
Bittner-2000	0.54 ± 0.08	0.53 ± 0.18	0.49 ± 0.16	0.71 ± 0.22
Chen-2002	0.91 ± 0.02	0.85 ± 0.07	0.81 ± 0.07	0.79 ± 0.12
Chowdary-2006	0.97 ± 0.01	0.97 ± 0.05	0.95 ± 0.05	0.94 ± 0.06
Golub-1999-v1	0.88 ± 0.02	0.86 ± 0.07	0.88 ± 0.08	0.90 ± 0.10
Lapointe-2004-v1	0.71 ± 0.06	0.78 ± 0.18	0.71 ± 0.14	0.63 ± 0.09
Liang-2005	0.89 ± 0.08	0.71 ± 0.14	0.76 ± 0.19	0.78 ± 0.18
Nutt-2003-v1	0.61 ± 0.10	0.54 ± 0.19	0.50 ± 0.17	0.40 ± 0.21
Nutt-2003-v2	0.78 ± 0.08	0.77 ± 0.21	0.87 ± 0.17	0.45 ± 0.27
Pomeroy-2002-v1	0.92 ± 0.05	0.84 ± 0.17	0.88 ± 0.16	0.73 ± 0.10
Risinger-2003	0.55 ± 0.08	0.52 ± 0.15	0.53 ± 0.19	0.59 ± 0.21
Shipp-2002-v1	0.87 ± 0.03	0.77 ± 0.12	0.84 ± 0.16	0.77 ± 0.05
Singh-2002	0.78 ± 0.03	0.74 ± 0.14	0.78 ± 0.12	0.69 ± 0.15
West-2001	0.92 ± 0.03	0.90 ± 0.11	0.84 ± 0.13	0.74 ± 0.19
Average rank	**1.44**	2.69	2.69	3.19

(b) *F-Measure results*

Data set	HEAD-DT	CART	C4.5	REP
Alizadeh-2000-v1	0.75 ± 0.05	0.67 ± 0.21	0.63 ± 0.26	0.72 ± 0.19
Alizadeh-2000-v2	0.91 ± 0.06	0.92 ± 0.10	0.84 ± 0.17	0.80 ± 0.19
Bhattacharjee-2001	0.94 ± 0.02	0.87 ± 0.12	0.89 ± 0.08	0.85 ± 0.07
Bittner-2000	0.53 ± 0.09	0.49 ± 0.21	0.45 ± 0.19	0.66 ± 0.28
Chen-2002	0.91 ± 0.02	0.85 ± 0.07	0.81 ± 0.07	0.78 ± 0.13
Chowdary-2006	0.97 ± 0.01	0.97 ± 0.05	0.95 ± 0.05	0.94 ± 0.07
Golub-1999-v1	0.88 ± 0.02	0.86 ± 0.07	0.87 ± 0.08	0.90 ± 0.10
Lapointe-2004-v1	0.69 ± 0.08	0.73 ± 0.20	0.68 ± 0.15	0.50 ± 0.13
Liang-2005	0.87 ± 0.10	0.67 ± 0.21	0.77 ± 0.22	0.71 ± 0.23
Nutt-2003-v1	0.60 ± 0.09	0.48 ± 0.19	0.46 ± 0.16	0.31 ± 0.22
Nutt-2003-v2	0.78 ± 0.08	0.73 ± 0.26	0.87 ± 0.17	0.34 ± 0.31
Pomeroy-2002-v1	0.92 ± 0.05	0.79 ± 0.22	0.85 ± 0.20	0.62 ± 0.14
Risinger-2003	0.53 ± 0.08	0.48 ± 0.19	0.53 ± 0.21	0.52 ± 0.25
Shipp-2002-v1	0.86 ± 0.03	0.75 ± 0.13	0.83 ± 0.17	0.70 ± 0.10
Singh-2002	0.78 ± 0.03	0.72 ± 0.18	0.77 ± 0.13	0.67 ± 0.18
West-2001	0.92 ± 0.03	0.89 ± 0.12	0.81 ± 0.17	0.71 ± 0.23
Average rank	**1.31**	2.69	2.63	3.38

Table 5.11 Results for the $\{7 \times 14\}$ configuration

(a) *Accuracy results*

Data set	HEAD-DT	CART	C4.5	REP
Alizadeh-2000-v1	0.79 ± 0.09	0.71 ± 0.16	0.68 ± 0.20	0.72 ± 0.19
Alizadeh-2000-v2	0.89 ± 0.05	0.92 ± 0.11	0.86 ± 0.15	0.84 ± 0.17
Bhattacharjee-2001	0.89 ± 0.03	0.89 ± 0.10	0.89 ± 0.08	0.86 ± 0.06
Chen-2002	0.84 ± 0.02	0.85 ± 0.07	0.81 ± 0.07	0.79 ± 0.12
Chowdary-2006	0.91 ± 0.04	0.97 ± 0.05	0.95 ± 0.05	0.94 ± 0.06
Golub-1999-v1	0.84 ± 0.03	0.86 ± 0.07	0.88 ± 0.08	0.90 ± 0.10
Lapointe-2004-v1	0.61 ± 0.09	0.78 ± 0.18	0.71 ± 0.14	0.63 ± 0.09
Liang-2005	0.87 ± 0.08	0.71 ± 0.14	0.76 ± 0.19	0.78 ± 0.18
Nutt-2003-v1	0.68 ± 0.07	0.54 ± 0.19	0.50 ± 0.17	0.40 ± 0.21
Nutt-2003-v2	0.76 ± 0.07	0.77 ± 0.21	0.87 ± 0.17	0.45 ± 0.27
Pomeroy-2002-v1	0.70 ± 0.12	0.84 ± 0.17	0.88 ± 0.16	0.73 ± 0.10
Shipp-2002-v1	0.86 ± 0.03	0.77 ± 0.12	0.84 ± 0.16	0.77 ± 0.05
Singh-2002	0.79 ± 0.02	0.74 ± 0.14	0.78 ± 0.12	0.69 ± 0.15
West-2001	0.80 ± 0.07	0.90 ± 0.11	0.84 ± 0.13	0.74 ± 0.19
Average rank	2.36	**2.14**	2.21	3.29

(b) *F-Measure results*

Data set	HEAD-DT	CART	C4.5	REP
Alizadeh-2000-v1	0.79 ± 0.09	0.67 ± 0.21	0.63 ± 0.26	0.72 ± 0.19
Alizadeh-2000-v2	0.91 ± 0.03	0.92 ± 0.10	0.84 ± 0.17	0.80 ± 0.19
Bhattacharjee-2001	0.89 ± 0.03	0.87 ± 0.12	0.89 ± 0.08	0.85 ± 0.07
Chen-2002	0.84 ± 0.02	0.85 ± 0.07	0.81 ± 0.07	0.78 ± 0.13
Chowdary-2006	0.91 ± 0.04	0.97 ± 0.05	0.95 ± 0.05	0.94 ± 0.07
Golub-1999-v1	0.83 ± 0.04	0.86 ± 0.07	0.87 ± 0.08	0.90 ± 0.10
Lapointe-2004-v1	0.60 ± 0.11	0.73 ± 0.20	0.68 ± 0.15	0.50 ± 0.13
Liang-2005	0.87 ± 0.08	0.67 ± 0.21	0.77 ± 0.22	0.71 ± 0.23
Nutt-2003-v1	0.65 ± 0.08	0.48 ± 0.19	0.46 ± 0.16	0.31 ± 0.22
Nutt-2003-v2	0.76 ± 0.07	0.73 ± 0.26	0.87 ± 0.17	0.34 ± 0.31
Pomeroy-2002-v1	0.69 ± 0.13	0.79 ± 0.22	0.85 ± 0.20	0.62 ± 0.14
Shipp-2002-v1	0.86 ± 0.03	0.75 ± 0.13	0.83 ± 0.17	0.70 ± 0.10
Singh-2002	0.79 ± 0.02	0.72 ± 0.18	0.77 ± 0.13	0.67 ± 0.18
West-2001	0.80 ± 0.07	0.89 ± 0.12	0.81 ± 0.17	0.71 ± 0.23
Average rank	**2.07**	2.21	2.21	3.50

in the second place and REPTree in the last position. The effect seen in Table 5.11a, in which HEAD-DT did not outperform the baseline methods, has a straightforward explanation: HEAD-DT optimises its generated algorithms according to the F-Measure evaluation measure. Since accuracy may be a misleading measure (it is not suitable for data sets with imbalanced class distributions), and several of the gene-expression data sets are imbalanced (i.e., they have a large difference in the

Table 5.12 Results for the {9 × 12} configuration

(a) *Accuracy results*

Data set	HEAD-DT	CART	C4.5	REP
Alizadeh-2000-v1	0.79 ± 0.09	0.71 ± 0.16	0.68 ± 0.20	0.72 ± 0.19
Alizadeh-2000-v2	0.89 ± 0.05	0.92 ± 0.11	0.86 ± 0.15	0.84 ± 0.17
Bhattacharjee-2001	0.90 ± 0.02	0.89 ± 0.10	0.89 ± 0.08	0.86 ± 0.06
Chen-2002	0.88 ± 0.03	0.85 ± 0.07	0.81 ± 0.07	0.79 ± 0.12
Golub-1999-v1	0.87 ± 0.02	0.86 ± 0.07	0.88 ± 0.08	0.90 ± 0.10
Lapointe-2004-v1	0.70 ± 0.05	0.78 ± 0.18	0.71 ± 0.14	0.63 ± 0.09
Liang-2005	0.88 ± 0.08	0.71 ± 0.14	0.76 ± 0.19	0.78 ± 0.18
Nutt-2003-v1	0.67 ± 0.09	0.54 ± 0.19	0.50 ± 0.17	0.40 ± 0.21
Nutt-2003-v2	0.77 ± 0.07	0.77 ± 0.21	0.87 ± 0.17	0.45 ± 0.27
Shipp-2002-v1	0.86 ± 0.03	0.84 ± 0.17	0.88 ± 0.16	0.73 ± 0.10
Pomeroy-2002-v1	0.92 ± 0.05	0.77 ± 0.12	0.84 ± 0.16	0.77 ± 0.05
Singh-2002	0.82 ± 0.04	0.74 ± 0.14	0.78 ± 0.12	0.69 ± 0.15
Average rank	**1.58**	2.67	2.33	3.42

(b) *F-Measure results*

Data set	HEAD-DT	CART	C4.5	REP
Alizadeh-2000-v1	0.78 ± 0.09	0.67 ± 0.21	0.63 ± 0.26	0.72 ± 0.19
Alizadeh-2000-v2	0.91 ± 0.03	0.92 ± 0.10	0.84 ± 0.17	0.80 ± 0.19
Bhattacharjee-2001	0.89 ± 0.02	0.87 ± 0.12	0.89 ± 0.08	0.85 ± 0.07
Chen-2002	0.88 ± 0.03	0.85 ± 0.07	0.81 ± 0.07	0.78 ± 0.13
Golub-1999-v1	0.87 ± 0.02	0.86 ± 0.07	0.87 ± 0.08	0.90 ± 0.10
Lapointe-2004-v1	0.68 ± 0.07	0.73 ± 0.20	0.68 ± 0.15	0.50 ± 0.13
Liang-2005	0.87 ± 0.09	0.67 ± 0.21	0.77 ± 0.22	0.71 ± 0.23
Nutt-2003-v1	0.66 ± 0.09	0.48 ± 0.19	0.46 ± 0.16	0.31 ± 0.22
Nutt-2003-v2	0.77 ± 0.07	0.73 ± 0.26	0.87 ± 0.17	0.34 ± 0.31
Shipp-2002-v1	0.86 ± 0.03	0.79 ± 0.22	0.85 ± 0.20	0.62 ± 0.14
Pomeroy-2002-v1	0.92 ± 0.05	0.75 ± 0.13	0.83 ± 0.17	0.70 ± 0.10
Singh-2002	0.82 ± 0.04	0.72 ± 0.18	0.77 ± 0.13	0.67 ± 0.18
Average rank	**1.42**	2.67	2.42	3.50

relative frequency of the most frequent and the least frequent classes in the data set), it seems fair to say that HEAD-DT also outperformed the baseline methods in configuration {7 × 14}, given that it generated algorithms whose F-Measure values are better than the values achieved by CART, C4.5, and REPTree.

Table 5.12 shows the results for configuration {9 × 12}. They are consistent to the previous configurations, in which HEAD-DT has the edge over the baseline methods, presenting the lowest average rank for both accuracy and F-Measure (1.58 and 1.42). C4.5 and CART presented similar ranks, whereas REPTree occupied the last position in the ranking.

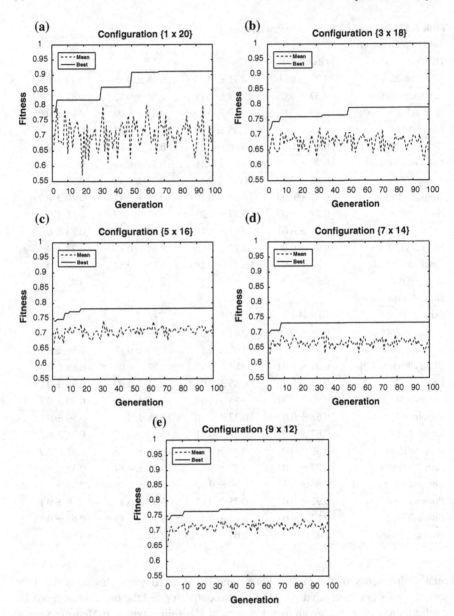

Fig. 5.1 Fitness evolution in HEAD-DT

Finally, Fig. 5.1 presents the fitness evolution in HEAD-DT across a full evolutionary cycle of 100 generations. We present both mean and best fitness of the population in a given generation, for all experimental configurations. Some interesting observations can be extracted from Fig. 5.1. First, when the meta-training set comprises a single data set (Fig. 5.1a), HEAD-DT is capable of continuously increase the fitness function value, and at the same time the population is reasonably heterogeneous (mean fitness value oscillates considerably). In the other extreme, when the meta-training set comprises 9 data sets (Fig. 5.1e), HEAD-DT has a harder time in optimising the fitness values, and the population is reasonably homogeneous (mean fitness value does not oscillate so much). The explanation for this behavior is that by increasing the number of data sets in the meta-training set, HEAD-DT has to find algorithms with a good performance trade-off within the meta-training set. In practice, modifications in the design of the algorithm that favor a given data set may very well harm another, and, hence, it is intuitively harder to design an algorithm that improves the performance of the generated decision trees in several data sets than in a single one. Conversely, it is also intuitive that a larger meta-training set leads to the design of a better "all-around" algorithm, i.e., an algorithm that is robust to the peculiarities of the data sets from the application domain.

5.2.1.3 Discussion

The experimental analysis conducted in the previous section aimed at comparing HEAD-DT to three baseline algorithms: CART, C4.5, and REPTree. We measured the performance of each algorithm according to accuracy and F-Measure, which are the most well-known criteria for evaluating classification algorithms. For verifying whether the number of data sets used in the meta-training set had an impact in the evolution of algorithms, we employed a consistent methodology that incrementally added random data sets from the set of available data, generating five different experimental configurations {#meta-training sets, #meta-test sets}: $\{1 \times 20\}$, $\{3 \times 18\}$, $\{5 \times 16\}$, $\{7 \times 14\}$, and $\{9 \times 12\}$. By analysing the average rank obtained by each method in the previously mentioned configurations, we conclude that:

- HEAD-DT is consistently the best-performing method, presenting the lowest average rank values among the four algorithms employed in the experimental analysis;
- C4.5 and CART's predictive performances are quite similar, which is consistent to the fact that both algorithms are still the state-of-the-art top-down decision-tree induction algorithms;
- REPTree, which is a variation of C4.5 that employs the reduced-error pruning strategy for simplifying the generated decision trees, is the worst-performing method in the group. Its disappointing results seem to indicate that the reduced-error pruning strategy is not particularly suited to the gene-expression data sets, probably because it requires an additional validation set. As previously observed, the gene-expression data sets have very few instances when compared to the currently giant databases from distinct application domains, and reducing their size for producing a validation set has certainly harmed REPTree's overall predictive performance.

For summarizing the average rank values obtained by each method in every experimental configuration, we gathered the rank values from Tables 5.8, 5.9, 5.10, 5.11 and 5.12 in Table 5.13. Values in bold indicate the best performing method according

Table 5.13 Summary of the experimental analysis regarding the homogeneous approach

Configuration	Rank	HEAD-DT	CART	C4.5	REPTree
$\{1 \times 20\}$	Accuracy	**1.75**	2.45	2.70	3.10
	F-measure	**1.55**	2.50	2.60	3.35
$\{3 \times 18\}$	Accuracy	**1.61**	2.61	2.72	3.05
	F-measure	**1.61**	2.61	2.56	3.22
$\{5 \times 16\}$	Accuracy	**1.44**	2.69	2.69	3.19
	F-measure	**1.31**	2.69	2.63	3.38
$\{7 \times 14\}$	Accuracy	2.36	**2.14**	2.21	3.29
	F-measure	**2.07**	2.21	2.21	3.50
$\{9 \times 12\}$	Accuracy	**1.58**	2.67	2.33	3.42
	F-measure	**1.42**	2.67	2.42	3.50
	Average	**1.67**	2.52	2.51	3.30

to the corresponding evaluation measure. These results show that HEAD-DT consistently presents the lowest average rank.

The last step of this empirical analysis is to verify whether the differences in rank values are statistically significant. For this particular analysis, we employ the graphical representation suggested by Demšar [7], the so-called *critical diagrams*. In this diagram, a horizontal line represents the axis on which we plot the average rank values of the methods. The axis is turned so that the lowest (best) ranks are to the right, since we perceive the methods on the right side as better. When comparing all the algorithms against each other, we connect the groups of algorithms that are not significantly different through a bold horizontal line. We also show the critical difference given by the Nemenyi test in the top of the graph.

Figure 5.2 shows the critical diagrams for all experimental configurations. Note that HEAD-DT outperforms C4.5, CART, and REPTree with statistical significance in configuration $\{5 \times 16\}$ for both accuracy (Fig. 5.2e) and F-Measure (Fig. 5.2f). The only scenarios in which there were no statistically significant differences between all methods are related to the accuracy measure in configuration $\{7 \times 14\}$ (Fig. 5.2g). The straightforward explanation for this case is that HEAD-DT optimises its solutions according to the F-Measure, even at the expense of accuracy. In the remaining scenarios, HEAD-DT always outperforms REPTree with statistical significance, which is not the case of CART and C4.5. In fact, CART and C4.5 are only able to outperform REPTree with statistical significance in the F-Measure evaluation in configuration $\{7 \times 14\}$ (Fig. 5.2h), suggesting once again that HEAD-DT should be preferred over any of the baseline methods.

As a final remark, considering the results of the four algorithms for all the 10 combinations of experimental configurations and performance measures as a whole, as summarized in Table 5.13, the decision-tree algorithm automatically designed by HEAD-DT obtained the best rank among the four algorithms in 9 out of the 10 rows of Table 5.13. Clearly, if the four algorithms had the same predictive performance (so that each algorithm would have a 25 % probability of being the winner), the probability that the algorithm designed by HEAD-DT would be the winner in 9 out

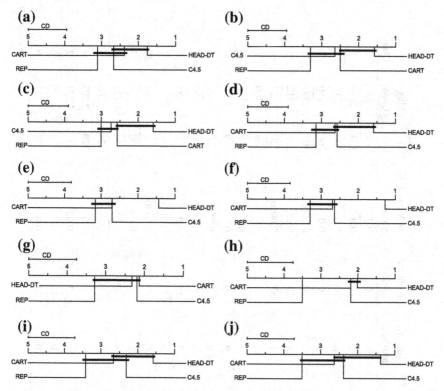

Fig. 5.2 Critical diagrams for the gene expression data. **a** Accuracy rank for $\{1 \times 20\}$. **b** F-Measure rank for $\{1 \times 20\}$. **c** Accuracy rank for $\{3 \times 18\}$. **d** F-Measure rank for $\{3 \times 18\}$. **e** Accuracy rank for $\{5 \times 16\}$. **f** F-Measure rank for $\{5 \times 16\}$. **g** Accuracy rank for $\{7 \times 14\}$. **h** F-Measure rank for $\{7 \times 14\}$. **i** Accuracy rank for $\{9 \times 12\}$. **j** F-Measure rank for $\{9 \times 12\}$

of 10 cases would be very small, so there is a high confidence that the results obtained by HEAD-DT are statistically valid as a whole.

5.2.2 The Heterogeneous Approach

In this set of experiments, we investigate the predictive performance of an automatica-lly-designed decision-tree induction algorithm tailored to a variety of distinct data sets. The goal is to evolve an algorithm capable of being robust across different data sets. For such, we make use of 67 publicly-available data sets from the UCI machine-learning repository[3] [8] (see Table 5.14). As in the homogeneous approach, we also randomly divided the 67 data sets into two groups: *parameter optimisation* and *experiments*. The 27 data sets in the *parameter optimisation* group are used for tuning the evolution parameters of HEAD-DT. The remaining 40 data sets from

---

[3] http://archive.ics.uci.edu/ml/.

Table 5.14 Summary of the 67 UCI data sets

	Data set	# Instances	# Attributes	Numeric attributes	Nominal attributes	% Missing Values	Min Class	Max Class	# Classes
Parameter optimisation	Balance-scale	625	3	4	0	0.0	49	288	3
	cmc	1,473	8	2	7	0.0	333	629	3
	Column-2C	310	5	6	0	0.0	100	210	2
	Column-3C	310	5	6	0	0.0	60	150	3
	Credit-a	690	14	6	9	0.0065	307	383	2
	Cylinder-bands	540	36	18	19	0.05	228	312	2
	Dermatology	366	33	1	33	0.0006	20	112	6
	Diabetes	768	7	8	0	0.0	268	500	2
	Ecoli	336	6	7	0	0.0	2	143	8
	Glass	214	8	9	0	0.0	9	76	6
	Hepatitis	155	18	6	13	0.0567	32	123	2
	Heart-statlog	270	12	13	0	0.0	120	150	2
	Lymph	148	17	3	15	0.0	2	81	4
	Mushroom	8,124	21	0	22	0.01387	3,916	4,208	2
	Primary-tumor	339	16	0	17	0.03904	1	84	21
	Segment	2,310	17	18	0	0.0	330	330	7
	Semeion	1,593	264	265	0	0.0	158	1,435	2
	Readings-2	5,456	1	2	0	0.0	328	2,205	4
	Readings-4	5,456	3	4	0	0.0	328	2,205	4
	Sick	3,772	26	6	21	0.0225	231	3,541	2
	Solar-flare-1	323	11	0	12	0.0	8	88	6
	Solar-flare-2	1,066	10	0	11	0.0	43	331	6

(continued)

Table 5.14 (continued)

	Data set	# Instances	# Attributes	Numeric attributes	Nominal attributes	% Missing Values	Min Class	Max Class	# Classes
	Sonar	208	59	60	0	0.0	97	111	2
	Sponge	76	43	0	44	0.00658	3	70	3
	Trains	10	25	0	26	0.1154	5	5	2
	Wine	178	12	13	0	0.0	48	71	3
	Zoo	101	16	1	16	0.0	4	41	7
Experiments	Abalone	4,177	7	7	1	0.0	1	689	28
	Anneal	898	37	6	32	0.0	8	684	5
	Arrhythmia	452	270	198	73	0.0033	2	245	13
	Audiology	226	68	0	69	0.02032	1	57	24
	Autos	205	24	15	10	0.0115	3	67	6
	Breast-cancer	286	8	0	9	0.0034	85	201	2
	Breast-w	699	8	9	0	0.0025	241	458	2
	Bridges1	105	10	3	8	0.0528	10	44	6
	Bridges2	105	10	0	11	0.0528	10	44	6
	Car	1,728	5	0	6	0.0	65	1,210	4
	Colic	368	21	7	15	0.2380	136	232	2
	Credit-g	1,000	19	7	13	0.0	300	700	2
	Flags	194	27	2	26	0.0	4	60	8
	Haberman	306	2	2	1	0.0	81	225	2
	Hayes-roth	160	3	4	0	0.0	31	65	3
	Heart-c	303	12	6	7	0.0018	138	165	2
	Heart-h	294	11	5	7	0.1391	106	188	2
	Ionosphere	351	32	33	0	0.0	126	225	2

(continued)

Table 5.14 (continued)

Data set	# Instances	# Attributes	Numeric attributes	Nominal attributes	% Missing Values	Min Class	Max Class	# Classes
Iris	150	3	4	0	0.0	50	50	3
kdd-synthetic	600	59	60	0	0.0	100	100	6
kr-vs-kp	3,196	35	0	36	0.0	1,527	1,669	2
Labor	57	15	8	8	0.3575	20	37	2
Liver-disorders	345	5	6	0	0.0	145	200	2
Lung-cancer	32	55	0	56	0.0028	9	13	3
Meta.data	528	20	20	1	0.0455	22	22	24
Morphological	2,000	5	6	0	0.0	200	200	10
mb-promoters	106	56	0	57	0.0	53	53	2
Postoperative-patient	90	7	0	8	0.0042	2	64	3
Shuttle-control	15	5	0	6	0.2889	6	9	2
Soybean	683	34	0	35	0.0978	8	92	19
Tae	151	4	3	2	0.0	49	52	3
Tempdiag	120	6	1	6	0.0	50	70	2
Tep.fea	3,572	6	7	0	0.0	303	1,733	3
Tic-tac-toe	958	8	0	9	0.0	332	626	2
Transfusion	748	3	4	0	0.0	178	570	2
Vehicle	846	17	18	0	0.0	199	218	4
Vote	435	15	0	16	0.0563	168	267	2
Vowel	990	12	10	3	0.0	90	90	11
Wine-red	1,599	10	11	0	0.0	10	681	6
Wine-white	4,898	10	11	0	0.0	5	2,198	7

the *experiments* group are used for evaluating the performance of the algorithm automatically designed by HEAD-DT.

Following the selection methodology previously presented, the 27 *parameter optimisation* data sets are arranged in 5 different experimental configurations {#training sets, #test sets}: $\{1 \times 26\}, \{3 \times 24\}, \{5 \times 22\}, \{7 \times 20\}$, and $\{9 \times 18\}$. Similarly, the 40 data sets in the *experiments* group are arranged in also 5 different experimental configurations {#training sets, #test sets}: $\{1 \times 39\}, \{3 \times 37\}, \{5 \times 35\}, \{7 \times 33\}$, and $\{9 \times 31\}$. Table 5.15 presents the randomly selected data sets according to the configurations detailed above.

5.2.2.1 Parameter Optimisation

Table 5.16 presents the results of the tuning experiments. We present the average ranking of each version of HEAD-DT (H-p) in the corresponding experimental configuration. For instance, when H-0.1 makes use of a single data set for evolving the optimal algorithm ($\{1 \times 26\}$), its predictive performance in the remaining 26 data sets gives H-0.1 the average rank position of 5.88 for the accuracy of its corresponding decision trees, and 5.96 for the F-Measure.

The Friedman and Nemenyi tests indicated statistically significant differences between the 9 distinct versions of HEAD-DT within each configuration. Depending on the configuration, different versions of HEAD-DT are outperformed by others with statistical significance. The only version of HEAD-DT that is not outperformed by any other version with statistical significance across all configurations is HEAD-0.3. Indeed, when we calculate the average of the average ranks for each H-p version across the distinct configurations and evaluation measures, HEAD-0.3 provides the lowest average of average ranks (3.99). Hence, HEAD-0.3 is the selected version to be executed over the 40 data sets in the *experiments* group, and it is henceforth referred simply as HEAD-DT.

5.2.2.2 Results

Tables 5.17, 5.18, 5.19, 5.20 and 5.21 show average values of accuracy and F-Measure obtained by HEAD-DT, CART, C4.5, and REPTree in configurations $\{1 \times 39\}$, $\{3 \times 37\}, \{5 \times 35\}, \{7 \times 33\}$, and $\{9 \times 31\}$. At the bottom of each table, we present the average rank position of each method. The lower the ranking, the better the method.

The first experiment is regarding configuration $\{1 \times 39\}$. Table 5.17 shows the result for this configuration considering accuracy (Table 5.17a) and F-Measure (Table 5.17b). Note that HEAD-DT and C4.5 are both the best performing method with respect to accuracy, with an average rank of 2.24, followed by CART (2.36) and REPTree (3.15). Considering the F-Measure, C4.5 has a slight edge over HEAD-DT (2.15 to 2.18), followed once again by CART (2.46) and REPTree (3.21).

Table 5.15 Meta-training and meta-test configurations for the UCI data

Data sets for parameter optimisation

{1 × 26}		{3 × 24}		{5 × 22}		{7 × 20}		{9 × 18}	
Meta-training	Meta-test	Meta-training	Meta-test	Meta-training	Meta-test	Meta-training	Meta-test	Meta-training	Meta-test
	Balance-scale		Balance-scale		Balance-scale		cmc		cmc
	cmc		cmc		cmc		Column_3C		column_3C
	Column_2C		Column_2C		Column_3C		Credit-a		Credit-a
	Column_3C		Column_3C		Credit-a		Cylinder-bands		Cylinder-bands
	Credit-a		Credit-a		Cylinder-bands		Dermatology		Dermatology
	Cylinder-bands		Cylinder-bands		Dermatology		Ecoli	Trains	Ecoli
	Dermatology		Dermatology		Ecoli	Trains	Heart-statlog	Zoo	Mushroom
	Ecoli		Ecoli	Column_2C	Heart-statlog	Zoo	Lymph	Wine	Diabetes
	Glass	Trains	Glass	Trains	Hepatitis	Wine	Mushroom	Column_2C	Primary-tumor
Trains	Heart-statlog	Zoo	Heart-statlog	Zoo	Lymph	Column_2C	Diabetes	Glass	Segment
	Hepatitis	Wine	Hepatitis	Wine	Mushroom	Glass	Primary-tumor	Balance-scale	Semeion
	Lymph		Lymph	Glass	Diabetes	Balance-scale	Segment	Hepatitis	Readings-2
	Mushroom		Mushroom		Primary-tumor	Hepatitis	Semeion	Heart-statlog	Readings-4
	Diabetes		Diabetes		Segment		Readings-2	Lymph	Sick
	Primary-tumor		Primary-tumor		Semeion		Readings-4		Solar-flare
	Segment		Segment		Readings-2		Sick		Solar-flare2
	Semeion		Semeion		Readings-4		Solar-flare		Sonar
	Readings-2		Readings-2		Sick		Solar-flare2		Sponge
	Readings-4		Readings-4		solar-flare		sonar		

(continued)

Table 5.15 (continued)

Data sets for parameter optimisation

{1 × 26}		{3 × 24}		{5 × 22}		{7 × 20}		{9 × 18}	
Meta-training	Meta-test	Meta-training	Meta-test	Meta-training	Meta-test	Meta-training	Meta-test	Meta-training	Meta-test
	Sick		Sick						
	Solar-flare		Solar-flare						
	Solar-flare2		Solar-flare2		Solar-flare2				
	Sonar		Sonar		Sonar				
	Sponge		Sponge		Sponge		Sponge		
	Wine								
	Zoo								

Data sets for the experimental analysis

{1 × 39}		{3 × 37}		{5 × 35}		{7 × 33}		{9 × 31}	
Meta-Training	Meta-Test	Meta-Training	Meta-Test	Meta-Training	Meta-Test	Meta-Training	Meta-Test	Meta-Training	Meta-Test
	Abalone		Abalone		Abalone		Abalone		Abalone
	Anneal		Anneal		Anneal		Anneal		Anneal
	Arrhythmia		Arrhythmia		Arrhythmia		Arrhythmia		Arrhythmia
	Audiology		Audiology		Audiology		Audiology		Audiology
	Autos		Autos		Autos		Autos		Autos
	Breast-cancer		Breast-cancer		Breast-cancer		Breast-cancer		Breast-cancer
	Breast-cancer		Breast-cancer		Breast-cancer		Breast-cancer		Breast-cancer
	Bridges1		Bridges1		Bridges1		Bridges2		Bridges2
Hayes	Bridges2		Bridges2	Hayes	Bridges2	Hayes	Car	Hayes	Car
								Labor	

(continued)

Table 5.15 (continued)

Data sets for the experimental analysis

{1×39} Meta-Training	{1×39} Meta-Test	{3×37} Meta-Training	{3×37} Meta-Test	{5×35} Meta-Training	{5×35} Meta-Test	{7×33} Meta-Training	{7×33} Meta-Test	{9×31} Meta-Training	{9×31} Meta-Test
						Bridges1		Bridges1	
	Car		Car		Car				
	Colic		Colic		Colic		Colic		Colic
	Credit-g		Credit-g		Credit-g		Credit-g		Credit-g
	Flags		Flags		Flags		Flags		Flags
	Haberman		Haberman	Haberman		Haberman		Haberman	
		Hayes		Hayes					
	Heart-c		Heart-c		Heart-c		Heart-c		Heart-c
	Heart-h		Heart-h		Heart-h		Heart-h		Heart-h
	Ionosphere		Ionosphere		Ionosphere		Ionosphere		Ionosphere
	Iris		Iris	Iris		Iris		Iris	
	kdd-synthetic		kdd-synthetic		kdd-synthetic		kdd-synthetic		kdd-synthetic
	kr-vs-kp		kr-vs-kp		kr-vs-kp		kr-vs-kp		kr-vs-kp
	Labor	Labor		Labor		Labor			
	Liver-disorders		Liver-disorders		Liver-disorders		Liver-disorders		Liver-disorders
	Lung-cancer		Lung-cancer		Lung-cancer		Lung-cancer	Lung-cancer	
	mb-promoters		mb-promoters		mb-promoters		mb-promoters		mb-promoters
	Meta.data		Meta.data		Meta.data		Meta.data		Meta.data
	Morphological		Morphological		Morphological		Morphological		Morphological
	Postoperative		Postoperative		Postoperative	Postoperative		Postoperative	
	Shuttle-control		Shuttle-control		Shuttle-control		Shuttle-control		Shuttle-control
	Soybean		Soybean		Soybean		Soybean		Soybean
	Tae	Tae		Tae		Tae		Tae	
			Tempdiag		Tempdiag		Tempdiag	Tempdiag	
			Tep.fea		Tep.fea		Tep.fea		Tep.fea
					Tic-tac-toe		Tic-tac-toe		Tic-tac-toe
					Transfusion		Transfusion		Transfusion
							Vehicle		Vehicle
							Vote		Vote
									Vowel
									Wine-red

(continued)

Table 5.15 (continued)

Data sets for the experimental analysis

	{1 × 39}		{3 × 37}		{5 × 35}		{7 × 33}		{9 × 31}
Meta-Training	Meta-Test	Meta-Training	Meta-Test	Meta-Training	Meta-Test	Meta-Training	Meta-Test	Meta-Training	Meta-Test
	Tempdiag								
	Tep.fea								
	Tic-tac-toe		Tic-tac-toe						
	Transfusion		Transfusion						
	Vehicle		Vehicle		Vehicle				
	Vote		Vote		Vote				
	Vowel		Vowel		Vowel		Vowel		
	Wine-red		Wine-red		Wine-red		Wine-red		
	Wine-white		Wine-white		Wine-white		Wine-white		Wine-white

Table 5.16 Results of the tuning experiments for the heterogeneous approach

Configuration	Rank	H-0.1	H-0.2	H-0.3	H-0.4	H-0.5	H-0.6	H-0.7	H-0.8	H-0.9
{1 × 26}	Accuracy	5.88	5.13	3.31	3.25	7.02	5.17	5.11	3.08	7.04
	F-measure	5.96	5.15	3.17	3.21	6.94	5.15	5.23	3.17	7.00
{3 × 24}	Accuracy	8.25	6.56	3.35	5.58	2.94	6.92	5.40	3.38	2.63
	F-measure	8.29	6.65	3.21	5.46	2.90	7.00	5.60	3.35	2.54
{5 × 22}	Accuracy	3.95	4.89	5.52	6.75	6.20	5.73	3.41	5.02	3.52
	F-measure	3.91	4.80	5.32	6.89	6.11	5.77	3.64	5.00	3.57
{7 × 20}	Accuracy	5.63	3.48	4.98	7.65	3.00	4.30	4.75	7.45	3.78
	F-measure	5.75	3.53	4.63	7.65	3.05	4.30	5.10	7.35	3.65
{9 × 18}	Accuracy	6.67	5.31	3.22	5.50	2.36	3.06	8.03	7.31	3.56
	F-measure	6.61	5.33	3.17	5.50	2.39	3.00	8.06	7.44	3.50
	Average	6.09	5.08	**3.99**	5.74	4.29	5.04	5.43	5.26	4.08

The next experiment concerns configuration {3 × 37}, whose results are presented in Table 5.18. In the accuracy analysis, C4.5 and CART are the first and second best-ranked algorithms, respectively, followed by HEAD-DT and REPTree. By looking at the F-Measure values, HEAD-DT replaces CART as the second best-ranked method. Note that this is not unusual, since HEAD-DT employs the average F-Measure as its fitness function, and thus optimises its algorithms in order for them to score the best possible F-Measure values. C4.5 keeps its place as the best-performing method, whereas REPTree is again the worst-performing method among the four algorithms, regardless of the evaluation measure.

Table 5.19 presents the experimental results for configuration {5 × 35}. The scenario is similar to the previous configuration, with C4.5 leading the ranking (average rank values of 2.13 for accuracy and 2.04 for F-Measure). C4.5 is followed by CART (2.19 and 2.27) and HEAD-DT (2.56 and 2.44). Finally, REPTree is in the bottom of the ranking, with average rank of 3.13 (accuracy) and 3.24 (F-Measure).

The experimental results for configuration {7 × 33} show a scenario in which HEAD-DT was capable of generating better algorithms than the remaining baseline methods. Table 5.20a, which depicts the accuracy values of each method, indicates that HEAD-DT leads the ranking with an average ranking position of 1.77—meaning its often the best-performing method in the 33 meta-test data sets. HEAD-DT is followed by C4.5 (2.35), CART (2.50), and REPTree (3.38). Table 5.20b shows that the F-Measure values provide the same ranking positions for all methods, with a clear edge to HEAD-DT (1.71), followed by C4.5 (2.29), CART (2.53), and REPTree (3.47).

Finally, Table 5.21 shows the results for configuration {9 × 31}. C4.5 returns to the top of the rank for both accuracy (2.03) and F-Measure (1.96) values. C4.5 is followed by CART (2.26 and 2.32), HEAD-DT (2.55 and 2.48), and REPTree (3.16 and 3.23).

Table 5.17 Results for the $\{1 \times 39\}$ configuration

(a) *Accuracy results*

Data set	HEAD	CART	C4.5	REP
Abalone	0.36 ± 0.10	0.26 ± 0.02	0.22 ± 0.02	0.26 ± 0.02
Anneal	0.92 ± 0.05	0.98 ± 0.01	0.99 ± 0.01	0.98 ± 0.02
Arrhythmia	0.66 ± 0.08	0.71 ± 0.05	0.66 ± 0.05	0.67 ± 0.06
Audiology	0.70 ± 0.06	0.74 ± 0.05	0.78 ± 0.07	0.74 ± 0.08
Autos	0.80 ± 0.01	0.78 ± 0.10	0.86 ± 0.06	0.65 ± 0.08
Breast-cancer	0.73 ± 0.02	0.69 ± 0.04	0.75 ± 0.08	0.69 ± 0.05
Bridges1	0.70 ± 0.03	0.53 ± 0.09	0.58 ± 0.11	0.40 ± 0.15
Bridges2	0.64 ± 0.07	0.54 ± 0.08	0.58 ± 0.13	0.40 ± 0.15
Car	0.86 ± 0.07	0.97 ± 0.02	0.93 ± 0.02	0.89 ± 0.02
Heart-c	0.83 ± 0.02	0.81 ± 0.04	0.77 ± 0.09	0.77 ± 0.08
Flags	0.71 ± 0.04	0.61 ± 0.10	0.63 ± 0.05	0.62 ± 0.10
Credit-g	0.78 ± 0.03	0.73 ± 0.04	0.71 ± 0.03	0.72 ± 0.06
Colic	0.74 ± 0.07	0.85 ± 0.08	0.86 ± 0.06	0.83 ± 0.06
Haberman	0.77 ± 0.01	0.75 ± 0.04	0.73 ± 0.09	0.73 ± 0.07
Heart-h	0.83 ± 0.02	0.77 ± 0.06	0.80 ± 0.08	0.80 ± 0.09
Ionosphere	0.90 ± 0.03	0.89 ± 0.03	0.89 ± 0.05	0.91 ± 0.02
Iris	0.96 ± 0.01	0.93 ± 0.05	0.94 ± 0.07	0.94 ± 0.05
kdd-synthetic	0.93 ± 0.03	0.88 ± 0.04	0.91 ± 0.04	0.88 ± 0.03
kr-vs-kp	0.91 ± 0.06	0.99 ± 0.01	0.99 ± 0.01	0.99 ± 0.01
Labor	0.79 ± 0.06	0.81 ± 0.17	0.79 ± 0.13	0.82 ± 0.21
Liver-disorders	0.77 ± 0.04	0.67 ± 0.09	0.67 ± 0.05	0.65 ± 0.05
Lung-cancer	0.66 ± 0.03	0.51 ± 0.33	0.45 ± 0.27	0.54 ± 0.17
Meta.data	0.11 ± 0.02	0.05 ± 0.03	0.04 ± 0.03	0.04 ± 0.00
Morphological	0.73 ± 0.03	0.72 ± 0.04	0.72 ± 0.02	0.72 ± 0.03
mb-promoters	0.78 ± 0.07	0.72 ± 0.14	0.80 ± 0.13	0.77 ± 0.15
Postoperative-patient	0.69 ± 0.01	0.71 ± 0.06	0.70 ± 0.05	0.69 ± 0.09
Shuttle-control	0.57 ± 0.04	0.65 ± 0.34	0.65 ± 0.34	0.65 ± 0.34
Soybean	0.69 ± 0.20	0.92 ± 0.04	0.92 ± 0.03	0.84 ± 0.05
Tae	0.64 ± 0.03	0.51 ± 0.12	0.60 ± 0.11	0.47 ± 0.12
Tempdiag	0.97 ± 0.04	1.00 ± 0.00	1.00 ± 0.00	1.00 ± 0.00
Tep.fea	0.65 ± 0.00	0.65 ± 0.02	0.65 ± 0.02	0.65 ± 0.02
Tic-tac-toe	0.83 ± 0.08	0.94 ± 0.02	0.86 ± 0.03	0.86 ± 0.03
Transfusion	0.80 ± 0.01	0.79 ± 0.03	0.78 ± 0.02	0.78 ± 0.02
Vehicle	0.78 ± 0.04	0.72 ± 0.04	0.74 ± 0.04	0.71 ± 0.04
Vote	0.95 ± 0.00	0.97 ± 0.02	0.97 ± 0.02	0.95 ± 0.03
Vowel	0.72 ± 0.16	0.82 ± 0.04	0.83 ± 0.03	0.70 ± 0.04
Wine-red	0.67 ± 0.06	0.63 ± 0.02	0.61 ± 0.03	0.60 ± 0.03

(continued)

Table 5.17 (continued)

(a) *Accuracy results*

Data set	HEAD	CART	C4.5	REP
Wine-white	0.62 ± 0.10	0.58 ± 0.02	0.61 ± 0.03	0.56 ± 0.02
Breast-w	0.94 ± 0.02	0.95 ± 0.02	0.95 ± 0.02	0.94 ± 0.03
Average rank	**2.24**	2.36	**2.24**	3.15

(b) *F-Measure results*

Data set	HEAD-DT	CART	C4.5	REP
Aabalone	0.34 ± 0.11	0.23 ± 0.02	0.21 ± 0.02	0.24 ± 0.02
Anneal	0.90 ± 0.08	0.98 ± 0.01	0.98 ± 0.01	0.98 ± 0.02
Arrhythmia	0.60 ± 0.11	0.67 ± 0.06	0.65 ± 0.06	0.63 ± 0.07
Audiology	0.66 ± 0.06	0.71 ± 0.05	0.75 ± 0.08	0.70 ± 0.09
Autos	0.80 ± 0.01	0.77 ± 0.10	0.85 ± 0.07	0.62 ± 0.07
Breast-cancer	0.71 ± 0.03	0.63 ± 0.05	0.70 ± 0.11	0.62 ± 0.06
Breast-w	0.94 ± 0.02	0.95 ± 0.02	0.95 ± 0.02	0.94 ± 0.03
Bridges1	0.68 ± 0.04	0.45 ± 0.06	0.52 ± 0.11	0.29 ± 0.11
Bridges2	0.62 ± 0.08	0.43 ± 0.05	0.51 ± 0.11	0.29 ± 0.11
Car	0.85 ± 0.08	0.97 ± 0.02	0.93 ± 0.02	0.89 ± 0.02
Heart-c	0.83 ± 0.02	0.80 ± 0.04	0.76 ± 0.09	0.77 ± 0.08
Flags	0.70 ± 0.05	0.57 ± 0.10	0.61 ± 0.05	0.58 ± 0.10
Credit-g	0.77 ± 0.04	0.71 ± 0.04	0.70 ± 0.02	0.70 ± 0.05
Colic	0.72 ± 0.09	0.84 ± 0.08	0.85 ± 0.07	0.83 ± 0.07
Haberman	0.75 ± 0.01	0.66 ± 0.06	0.69 ± 0.10	0.68 ± 0.08
Heart-h	0.83 ± 0.02	0.76 ± 0.06	0.80 ± 0.07	0.79 ± 0.09
Ionosphere	0.90 ± 0.04	0.89 ± 0.03	0.88 ± 0.05	0.91 ± 0.02
Iris	0.96 ± 0.01	0.93 ± 0.06	0.94 ± 0.07	0.94 ± 0.05
kdd-synthetic	0.93 ± 0.03	0.88 ± 0.04	0.91 ± 0.04	0.87 ± 0.04
kr-vs-kp	0.91 ± 0.06	0.99 ± 0.01	0.99 ± 0.01	0.99 ± 0.01
Labor	0.76 ± 0.09	0.80 ± 0.17	0.78 ± 0.12	0.82 ± 0.21
Liver-disorders	0.77 ± 0.04	0.66 ± 0.09	0.66 ± 0.05	0.63 ± 0.05
Lung-cancer	0.65 ± 0.04	0.42 ± 0.32	0.35 ± 0.29	0.42 ± 0.19
Meta.data	0.10 ± 0.03	0.02 ± 0.01	0.02 ± 0.02	0.00 ± 0.00
Morphological	0.71 ± 0.04	0.70 ± 0.04	0.70 ± 0.02	0.70 ± 0.03
mb-promoters	0.78 ± 0.07	0.71 ± 0.14	0.79 ± 0.14	0.76 ± 0.15
Postoperative-patient	0.64 ± 0.02	0.59 ± 0.08	0.59 ± 0.07	0.58 ± 0.09
Shuttle-control	0.55 ± 0.06	0.57 ± 0.39	0.57 ± 0.39	0.57 ± 0.39
Soybean	0.66 ± 0.21	0.91 ± 0.04	0.92 ± 0.04	0.82 ± 0.06
Tae	0.64 ± 0.03	0.49 ± 0.15	0.59 ± 0.12	0.45 ± 0.12
Tempdiag	0.97 ± 0.04	1.00 ± 0.00	1.00 ± 0.00	1.00 ± 0.00
Tep.fea	0.61 ± 0.00	0.61 ± 0.02	0.61 ± 0.02	0.61 ± 0.02

(continued)

Table 5.17 (continued)

(b) *F-Measure results*

Data set	HEAD-DT	CART	C4.5	REP
Tic-tac-toe	0.83 ± 0.09	0.94 ± 0.02	0.86 ± 0.03	0.86 ± 0.03
Transfusion	0.78 ± 0.01	0.76 ± 0.03	0.77 ± 0.03	0.76 ± 0.02
Vehicle	0.77 ± 0.04	0.71 ± 0.05	0.73 ± 0.04	0.70 ± 0.04
Vote	0.95 ± 0.00	0.97 ± 0.02	0.97 ± 0.02	0.95 ± 0.03
Vowel	0.71 ± 0.17	0.82 ± 0.04	0.83 ± 0.03	0.70 ± 0.04
Wine-red	0.66 ± 0.07	0.61 ± 0.02	0.61 ± 0.03	0.58 ± 0.03
Wine-white	0.60 ± 0.12	0.58 ± 0.03	0.60 ± 0.02	0.55 ± 0.02
Average rank	2.18	2.46	**2.15**	3.21

5.2.2.3 Discussion

The experimental analysis conducted in the previous section aimed at comparing the performance of HEAD-DT when it is employed to generate an "all-around" decision-tree algorithm—i.e., an algorithm capable of performing reasonably well in a variety of distinct data sets. We once again measured the performance of each algorithm according to accuracy and F-Measure, which are the most well-known criteria for evaluating classification algorithms. In order to verify whether the number of data sets used in the meta-training set had an impact in the evolution of algorithms, we employed the same methodology as in the homogeneous approach, resulting in five different experimental configurations {#meta-training sets, #meta-test sets}: $\{1 \times 39\}, \{3 \times 37\}, \{5 \times 35\}, \{7 \times 33\}$, and $\{9 \times 31\}$. By analysing the average rank obtained by each method in the previously mentioned configurations, we conclude that:

- C4.5 is the best-performing method, consistently presenting the lowest average rank values among the four algorithms employed in the experimental analysis;
- HEAD-DT is capable of generating competitive algorithms, and eventually the best-performing algorithm (e.g., configuration $\{7 \times 33\}$). Since its performance is heavily dependent on the data sets that comprise the meta-training set, it is to be expected that HEAD-DT eventually generates algorithms that are too-specific for the meta-training set (it suffers from a kind of "meta-overfitting"). Unfortunately, there is no easy solution to avoid this type of overfitting (see Sect. 5.2.3 for more details on this matter);
- REPTree is the worst-performing method in the group. This is expected given that its reduced-error pruning strategy is said to be effective only for very large data sets, considering that it requires a validation set.

For summarizing the average rank values obtained by each method in every experimental configuration, we gathered the rank values from Tables 5.17, 5.18, 5.19, 5.20 and 5.21 in Table 5.22. Values in bold indicate the best performing method according to the corresponding evaluation measure.

Table 5.18 Results for the $\{3 \times 37\}$ configuration

(a) *Accuracy results*

Data set	HEAD	CART	C4.5	REP
Abalone	0.27 ± 0.00	0.26 ± 0.02	0.22 ± 0.02	0.26 ± 0.02
Anneal	0.98 ± 0.01	0.98 ± 0.01	0.99 ± 0.01	0.98 ± 0.02
Arrhythmia	0.66 ± 0.10	0.71 ± 0.05	0.66 ± 0.05	0.67 ± 0.06
Audiology	0.73 ± 0.02	0.74 ± 0.05	0.78 ± 0.07	0.74 ± 0.08
Autos	0.73 ± 0.08	0.78 ± 0.10	0.86 ± 0.06	0.65 ± 0.08
Breast-cancer	0.74 ± 0.01	0.69 ± 0.04	0.75 ± 0.08	0.69 ± 0.05
Breast-w	0.94 ± 0.01	0.95 ± 0.02	0.95 ± 0.02	0.94 ± 0.03
Bridges1	0.70 ± 0.06	0.53 ± 0.09	0.58 ± 0.11	0.40 ± 0.15
Bridges2	0.69 ± 0.05	0.54 ± 0.08	0.58 ± 0.13	0.40 ± 0.15
Car	0.84 ± 0.02	0.97 ± 0.02	0.93 ± 0.02	0.89 ± 0.02
Heart-c	0.81 ± 0.00	0.81 ± 0.04	0.77 ± 0.09	0.77 ± 0.08
Flags	0.69 ± 0.01	0.61 ± 0.10	0.63 ± 0.05	0.62 ± 0.10
Credit-g	0.74 ± 0.00	0.73 ± 0.04	0.71 ± 0.03	0.72 ± 0.06
Colic	0.78 ± 0.12	0.85 ± 0.08	0.86 ± 0.06	0.83 ± 0.06
Haberman	0.77 ± 0.00	0.75 ± 0.04	0.73 ± 0.09	0.73 ± 0.07
Heart-h	0.80 ± 0.01	0.77 ± 0.06	0.80 ± 0.08	0.80 ± 0.09
Ionosphere	0.92 ± 0.03	0.89 ± 0.03	0.89 ± 0.05	0.91 ± 0.02
Iris	0.96 ± 0.00	0.93 ± 0.05	0.94 ± 0.07	0.94 ± 0.05
kdd-synthetic	0.95 ± 0.01	0.88 ± 0.04	0.91 ± 0.04	0.88 ± 0.03
kr-vs-kp	0.91 ± 0.03	0.99 ± 0.01	0.99 ± 0.01	0.99 ± 0.01
Liver-disorders	0.73 ± 0.01	0.67 ± 0.09	0.67 ± 0.05	0.65 ± 0.05
Lung-cancer	0.69 ± 0.00	0.51 ± 0.33	0.45 ± 0.27	0.54 ± 0.17
Meta.data	0.08 ± 0.04	0.05 ± 0.03	0.04 ± 0.03	0.04 ± 0.00
Morphological	0.71 ± 0.00	0.72 ± 0.04	0.72 ± 0.02	0.72 ± 0.03
mb-promoters	0.88 ± 0.02	0.72 ± 0.14	0.80 ± 0.13	0.77 ± 0.15
Postoperative-patient	0.70 ± 0.02	0.71 ± 0.06	0.70 ± 0.05	0.69 ± 0.09
Shuttle-control	0.60 ± 0.02	0.65 ± 0.34	0.65 ± 0.34	0.65 ± 0.34
Soybean	0.79 ± 0.06	0.92 ± 0.04	0.92 ± 0.03	0.84 ± 0.05
Tempdiag	1.00 ± 0.00	1.00 ± 0.00	1.00 ± 0.00	1.00 ± 0.00
Tep.fea	0.65 ± 0.00	0.65 ± 0.02	0.65 ± 0.02	0.65 ± 0.02
Tic-tac-toe	0.76 ± 0.04	0.94 ± 0.02	0.86 ± 0.03	0.86 ± 0.03
Transfusion	0.79 ± 0.01	0.79 ± 0.03	0.78 ± 0.02	0.78 ± 0.02
Vehicle	0.74 ± 0.00	0.72 ± 0.04	0.74 ± 0.04	0.71 ± 0.04
Vote	0.95 ± 0.01	0.97 ± 0.02	0.97 ± 0.02	0.95 ± 0.03
Vowel	0.59 ± 0.08	0.82 ± 0.04	0.83 ± 0.03	0.70 ± 0.04
Wine-red	0.59 ± 0.01	0.63 ± 0.02	0.61 ± 0.03	0.60 ± 0.03
Wine-white	0.52 ± 0.01	0.58 ± 0.02	0.61 ± 0.03	0.56 ± 0.02
Average rank	2.45	2.26	**2.19**	3.11

(continued)

Table 5.18 (continued)

(b) *F-Measure results*

Data set	HEAD	CART	C4.5	REP
Abalone	0.23 ± 0.00	0.23 ± 0.02	0.21 ± 0.02	0.24 ± 0.02
Anneal	0.97 ± 0.01	0.98 ± 0.01	0.98 ± 0.01	0.98 ± 0.02
Arrhythmia	0.58 ± 0.17	0.67 ± 0.06	0.65 ± 0.06	0.63 ± 0.07
Audiology	0.70 ± 0.02	0.71 ± 0.05	0.75 ± 0.08	0.70 ± 0.09
Autos	0.73 ± 0.08	0.77 ± 0.10	0.85 ± 0.07	0.62 ± 0.07
Breast-cancer	0.72 ± 0.00	0.63 ± 0.05	0.70 ± 0.11	0.62 ± 0.06
Breast-w	0.94 ± 0.01	0.95 ± 0.02	0.95 ± 0.02	0.94 ± 0.03
Bridges1	0.68 ± 0.06	0.45 ± 0.06	0.52 ± 0.11	0.29 ± 0.11
Bridges2	0.68 ± 0.06	0.43 ± 0.05	0.51 ± 0.11	0.29 ± 0.11
Car	0.83 ± 0.02	0.97 ± 0.02	0.93 ± 0.02	0.89 ± 0.02
Heart-c	0.81 ± 0.00	0.80 ± 0.04	0.76 ± 0.09	0.77 ± 0.08
Flags	0.68 ± 0.01	0.57 ± 0.10	0.61 ± 0.05	0.58 ± 0.10
Credit-g	0.73 ± 0.00	0.71 ± 0.04	0.70 ± 0.02	0.70 ± 0.05
Colic	0.72 ± 0.18	0.84 ± 0.08	0.85 ± 0.07	0.83 ± 0.07
Haberman	0.75 ± 0.01	0.66 ± 0.06	0.69 ± 0.10	0.68 ± 0.08
Heart-h	0.80 ± 0.01	0.76 ± 0.06	0.80 ± 0.07	0.79 ± 0.09
Ionosphere	0.92 ± 0.03	0.89 ± 0.03	0.88 ± 0.05	0.91 ± 0.02
Iris	0.96 ± 0.00	0.93 ± 0.06	0.94 ± 0.07	0.94 ± 0.05
kdd-synthetic	0.95 ± 0.01	0.88 ± 0.04	0.91 ± 0.04	0.87 ± 0.04
kr-vs-kp	0.91 ± 0.03	0.99 ± 0.01	0.99 ± 0.01	0.99 ± 0.01
Liver-disorders	0.72 ± 0.02	0.66 ± 0.09	0.66 ± 0.05	0.63 ± 0.05
Lung-cancer	0.69 ± 0.00	0.42 ± 0.32	0.35 ± 0.29	0.42 ± 0.19
Meta.data	0.06 ± 0.03	0.02 ± 0.01	0.02 ± 0.02	0.00 ± 0.00
Morphological	0.70 ± 0.00	0.70 ± 0.04	0.70 ± 0.02	0.70 ± 0.03
mb-promoters	0.88 ± 0.02	0.71 ± 0.14	0.79 ± 0.14	0.76 ± 0.15
Postoperative-patient	0.67 ± 0.02	0.59 ± 0.08	0.59 ± 0.07	0.58 ± 0.09
Shuttle-control	0.58 ± 0.02	0.57 ± 0.39	0.57 ± 0.39	0.57 ± 0.39
Soybean	0.76 ± 0.07	0.91 ± 0.04	0.92 ± 0.04	0.82 ± 0.06
Tempdiag	1.00 ± 0.00	1.00 ± 0.00	1.00 ± 0.00	1.00 ± 0.00
Tep.fea	0.61 ± 0.00	0.61 ± 0.02	0.61 ± 0.02	0.61 ± 0.02
Tic-tac-toe	0.76 ± 0.04	0.94 ± 0.02	0.86 ± 0.03	0.86 ± 0.03
Transfusion	0.77 ± 0.00	0.76 ± 0.03	0.77 ± 0.03	0.76 ± 0.02
Vehicle	0.74 ± 0.00	0.71 ± 0.05	0.73 ± 0.04	0.70 ± 0.04
Vote	0.95 ± 0.01	0.97 ± 0.02	0.97 ± 0.02	0.95 ± 0.03
Vowel	0.58 ± 0.09	0.82 ± 0.04	0.83 ± 0.03	0.70 ± 0.04
Wine-red	0.57 ± 0.01	0.61 ± 0.02	0.61 ± 0.03	0.58 ± 0.03
Wine-white	0.48 ± 0.02	0.58 ± 0.03	0.60 ± 0.02	0.55 ± 0.02
Average rank	2.23	2.42	**2.15**	3.20

Table 5.19 Results for the $\{5 \times 35\}$ configuration

(a) *Accuracy results*

Data set	HEAD	CART	C4.5	REP
Abalone	0.27 ± 0.00	0.26 ± 0.02	0.22 ± 0.02	0.26 ± 0.02
Anneal	0.97 ± 0.00	0.98 ± 0.01	0.99 ± 0.01	0.98 ± 0.02
Arrhythmia	0.58 ± 0.08	0.71 ± 0.05	0.66 ± 0.05	0.67 ± 0.06
Audiology	0.76 ± 0.00	0.74 ± 0.05	0.78 ± 0.07	0.74 ± 0.08
Autos	0.67 ± 0.06	0.78 ± 0.10	0.86 ± 0.06	0.65 ± 0.08
Breast-cancer	0.75 ± 0.00	0.69 ± 0.04	0.75 ± 0.08	0.69 ± 0.05
Breast-w	0.93 ± 0.00	0.95 ± 0.02	0.95 ± 0.02	0.94 ± 0.03
Bridges1	0.64 ± 0.03	0.53 ± 0.09	0.58 ± 0.11	0.40 ± 0.15
Bridges2	0.64 ± 0.03	0.54 ± 0.08	0.58 ± 0.13	0.40 ± 0.15
Car	0.82 ± 0.02	0.97 ± 0.02	0.93 ± 0.02	0.89 ± 0.02
Heart-c	0.81 ± 0.01	0.81 ± 0.04	0.77 ± 0.09	0.77 ± 0.08
Flags	0.68 ± 0.01	0.61 ± 0.10	0.63 ± 0.05	0.62 ± 0.10
Credit-g	0.75 ± 0.00	0.73 ± 0.04	0.71 ± 0.03	0.72 ± 0.06
Colic	0.68 ± 0.09	0.85 ± 0.08	0.86 ± 0.06	0.83 ± 0.06
Heart-h	0.77 ± 0.05	0.77 ± 0.06	0.80 ± 0.08	0.80 ± 0.09
Ionosphere	0.89 ± 0.00	0.89 ± 0.03	0.89 ± 0.05	0.91 ± 0.02
kdd-synthetic	0.96 ± 0.00	0.88 ± 0.04	0.91 ± 0.04	0.88 ± 0.03
kr-vs-kp	0.95 ± 0.00	0.99 ± 0.01	0.99 ± 0.01	0.99 ± 0.01
Liver-disorders	0.74 ± 0.01	0.67 ± 0.09	0.67 ± 0.05	0.65 ± 0.05
Lung-cancer	0.69 ± 0.00	0.51 ± 0.33	0.45 ± 0.27	0.54 ± 0.17
Meta.data	0.04 ± 0.02	0.05 ± 0.03	0.04 ± 0.03	0.04 ± 0.00
Morphological	0.70 ± 0.00	0.72 ± 0.04	0.72 ± 0.02	0.72 ± 0.03
mb-promoters	0.86 ± 0.01	0.72 ± 0.14	0.80 ± 0.13	0.77 ± 0.15
Postoperative-patient	0.72 ± 0.02	0.71 ± 0.06	0.70 ± 0.05	0.69 ± 0.09
Shuttle-control	0.61 ± 0.01	0.65 ± 0.34	0.65 ± 0.34	0.65 ± 0.34
Soybean	0.72 ± 0.02	0.92 ± 0.04	0.92 ± 0.03	0.84 ± 0.05
Tempdiag	1.00 ± 0.00	1.00 ± 0.00	1.00 ± 0.00	1.00 ± 0.00
Tep.fea	0.65 ± 0.00	0.65 ± 0.02	0.65 ± 0.02	0.65 ± 0.02
Tic-tac-toe	0.73 ± 0.03	0.94 ± 0.02	0.86 ± 0.03	0.86 ± 0.03
Transfusion	0.79 ± 0.00	0.79 ± 0.03	0.78 ± 0.02	0.78 ± 0.02
Vehicle	0.74 ± 0.00	0.72 ± 0.04	0.74 ± 0.04	0.71 ± 0.04
Vote	0.96 ± 0.00	0.97 ± 0.02	0.97 ± 0.02	0.95 ± 0.03
Vowel	0.50 ± 0.01	0.82 ± 0.04	0.83 ± 0.03	0.70 ± 0.04
Wine-red	0.60 ± 0.00	0.63 ± 0.02	0.61 ± 0.03	0.60 ± 0.03
Wine-white	0.54 ± 0.00	0.58 ± 0.02	0.61 ± 0.03	0.56 ± 0.02
Average rank	2.56	2.19	**2.13**	3.13

(continued)

Table 5.19 (continued)

(b) *F-Measure results*

Data set	HEAD	CART	C4.5	REP
Abalone	0.24 ± 0.00	0.23 ± 0.02	0.21 ± 0.02	0.24 ± 0.02
Anneal	0.97 ± 0.00	0.98 ± 0.01	0.98 ± 0.01	0.98 ± 0.02
Arrhythmia	0.45 ± 0.13	0.67 ± 0.06	0.65 ± 0.06	0.63 ± 0.07
Audiology	0.73 ± 0.00	0.71 ± 0.05	0.75 ± 0.08	0.70 ± 0.09
Autos	0.67 ± 0.06	0.77 ± 0.10	0.85 ± 0.07	0.62 ± 0.07
Breast-cancer	0.73 ± 0.00	0.63 ± 0.05	0.70 ± 0.11	0.62 ± 0.06
Breast-w	0.93 ± 0.00	0.95 ± 0.02	0.95 ± 0.02	0.94 ± 0.03
Bridges1	0.63 ± 0.03	0.45 ± 0.06	0.52 ± 0.11	0.29 ± 0.11
Bridges2	0.63 ± 0.04	0.43 ± 0.05	0.51 ± 0.11	0.29 ± 0.11
Car	0.81 ± 0.02	0.97 ± 0.02	0.93 ± 0.02	0.89 ± 0.02
Heart-c	0.80 ± 0.01	0.80 ± 0.04	0.76 ± 0.09	0.77 ± 0.08
Flags	0.67 ± 0.01	0.57 ± 0.10	0.61 ± 0.05	0.58 ± 0.10
Credit-g	0.73 ± 0.00	0.71 ± 0.04	0.70 ± 0.02	0.70 ± 0.05
Colic	0.57 ± 0.14	0.84 ± 0.08	0.85 ± 0.07	0.83 ± 0.07
Heart-h	0.74 ± 0.07	0.76 ± 0.06	0.80 ± 0.07	0.79 ± 0.09
Ionosphere	0.89 ± 0.01	0.89 ± 0.03	0.88 ± 0.05	0.91 ± 0.02
kdd-synthetic	0.96 ± 0.00	0.88 ± 0.04	0.91 ± 0.04	0.87 ± 0.04
kr-vs-kp	0.95 ± 0.00	0.99 ± 0.01	0.99 ± 0.01	0.99 ± 0.01
Liver-disorders	0.73 ± 0.01	0.66 ± 0.09	0.66 ± 0.05	0.63 ± 0.05
Lung-cancer	0.69 ± 0.00	0.42 ± 0.32	0.35 ± 0.29	0.42 ± 0.19
Meta.data	0.02 ± 0.01	0.02 ± 0.01	0.02 ± 0.02	0.00 ± 0.00
Morphological	0.69 ± 0.01	0.70 ± 0.04	0.70 ± 0.02	0.70 ± 0.03
mb-promoters	0.86 ± 0.01	0.71 ± 0.14	0.79 ± 0.14	0.76 ± 0.15
Postoperative-patient	0.69 ± 0.03	0.59 ± 0.08	0.59 ± 0.07	0.58 ± 0.09
Shuttle-control	0.57 ± 0.02	0.57 ± 0.39	0.57 ± 0.39	0.57 ± 0.39
Soybean	0.68 ± 0.01	0.91 ± 0.04	0.92 ± 0.04	0.82 ± 0.06
Tempdiag	1.00 ± 0.00	1.00 ± 0.00	1.00 ± 0.00	1.00 ± 0.00
Tep.fea	0.61 ± 0.00	0.61 ± 0.02	0.61 ± 0.02	0.61 ± 0.02
Tic-tac-toe	0.72 ± 0.04	0.94 ± 0.02	0.86 ± 0.03	0.86 ± 0.03
Transfusion	0.77 ± 0.00	0.76 ± 0.03	0.77 ± 0.03	0.76 ± 0.02
Vehicle	0.74 ± 0.00	0.71 ± 0.05	0.73 ± 0.04	0.70 ± 0.04
Vote	0.96 ± 0.00	0.97 ± 0.02	0.97 ± 0.02	0.95 ± 0.03
Vowel	0.48 ± 0.02	0.82 ± 0.04	0.83 ± 0.03	0.70 ± 0.04
Wine-red	0.59 ± 0.00	0.61 ± 0.02	0.61 ± 0.03	0.58 ± 0.03
Wine-white	0.51 ± 0.00	0.58 ± 0.03	0.60 ± 0.02	0.55 ± 0.02
Average rank	2.44	2.27	**2.04**	3.24

Table 5.20 Results for the $\{7 \times 33\}$ configuration

(a) *Accuracy results*

Data set	HEAD	CART	C4.5	REP
Abalone	0.29 ± 0.01	0.26 ± 0.02	0.22 ± 0.02	0.26 ± 0.02
Anneal	0.98 ± 0.01	0.98 ± 0.01	0.99 ± 0.01	0.98 ± 0.02
Arrhythmia	0.78 ± 0.02	0.71 ± 0.05	0.66 ± 0.05	0.67 ± 0.06
Audiology	0.79 ± 0.00	0.74 ± 0.05	0.78 ± 0.07	0.74 ± 0.08
Autos	0.84 ± 0.04	0.78 ± 0.10	0.86 ± 0.06	0.65 ± 0.08
Breast-cancer	0.75 ± 0.00	0.69 ± 0.04	0.75 ± 0.08	0.69 ± 0.05
Breast-w	0.96 ± 0.01	0.95 ± 0.02	0.95 ± 0.02	0.94 ± 0.03
Bridges2	0.71 ± 0.01	0.54 ± 0.08	0.58 ± 0.13	0.40 ± 0.15
Car	0.92 ± 0.02	0.97 ± 0.02	0.93 ± 0.02	0.89 ± 0.02
Heart-c	0.83 ± 0.01	0.81 ± 0.04	0.77 ± 0.09	0.77 ± 0.08
Flags	0.73 ± 0.01	0.61 ± 0.10	0.63 ± 0.05	0.62 ± 0.10
Credit-g	0.76 ± 0.00	0.73 ± 0.04	0.71 ± 0.03	0.72 ± 0.06
Colic	0.88 ± 0.01	0.85 ± 0.08	0.86 ± 0.06	0.83 ± 0.06
Heart-h	0.83 ± 0.01	0.77 ± 0.06	0.80 ± 0.08	0.80 ± 0.09
Ionosphere	0.94 ± 0.01	0.89 ± 0.03	0.89 ± 0.05	0.91 ± 0.02
kdd-synthetic	0.95 ± 0.00	0.88 ± 0.04	0.91 ± 0.04	0.88 ± 0.03
kr-vs-kp	0.96 ± 0.00	0.99 ± 0.01	0.99 ± 0.01	0.99 ± 0.01
Liver-disorders	0.75 ± 0.00	0.67 ± 0.09	0.67 ± 0.05	0.65 ± 0.05
Lung-cancer	0.71 ± 0.02	0.51 ± 0.33	0.45 ± 0.27	0.54 ± 0.17
Meta.data	0.14 ± 0.02	0.05 ± 0.03	0.04 ± 0.03	0.04 ± 0.00
Morphological	0.74 ± 0.01	0.72 ± 0.04	0.72 ± 0.02	0.72 ± 0.03
mb-promoters	0.89 ± 0.01	0.72 ± 0.14	0.80 ± 0.13	0.77 ± 0.15
Shuttle-control	0.59 ± 0.02	0.65 ± 0.34	0.65 ± 0.34	0.65 ± 0.34
Soybean	0.82 ± 0.08	0.92 ± 0.04	0.92 ± 0.03	0.84 ± 0.05
Tempdiag	1.00 ± 0.00	1.00 ± 0.00	1.00 ± 0.00	1.00 ± 0.00
Tep.fea	0.65 ± 0.00	0.65 ± 0.02	0.65 ± 0.02	0.65 ± 0.02
Tic-tac-toe	0.94 ± 0.02	0.94 ± 0.02	0.86 ± 0.03	0.86 ± 0.03
Transfusion	0.79 ± 0.00	0.79 ± 0.03	0.78 ± 0.02	0.78 ± 0.02
Vehicle	0.77 ± 0.01	0.72 ± 0.04	0.74 ± 0.04	0.71 ± 0.04
Vote	0.96 ± 0.00	0.97 ± 0.02	0.97 ± 0.02	0.95 ± 0.03
Vowel	0.76 ± 0.07	0.82 ± 0.04	0.83 ± 0.03	0.70 ± 0.04
Wine-red	0.64 ± 0.01	0.63 ± 0.02	0.61 ± 0.03	0.60 ± 0.03
Wine-white	0.55 ± 0.01	0.58 ± 0.02	0.61 ± 0.03	0.56 ± 0.02
Average rank	**1.77**	2.50	2.35	3.38

(continued)

Table 5.20 (continued)

(b) *F-Measure results*

Data set	HEAD	CART	C4.5	REP
Abalone	0.27 ± 0.01	0.23 ± 0.02	0.21 ± 0.02	0.24 ± 0.02
Anneal	0.98 ± 0.01	0.98 ± 0.01	0.98 ± 0.01	0.98 ± 0.02
Arrhythmia	0.76 ± 0.02	0.67 ± 0.06	0.65 ± 0.06	0.63 ± 0.07
Audiology	0.77 ± 0.01	0.71 ± 0.05	0.75 ± 0.08	0.70 ± 0.09
Autos	0.85 ± 0.04	0.77 ± 0.10	0.85 ± 0.07	0.62 ± 0.07
Breast-cancer	0.73 ± 0.00	0.63 ± 0.05	0.70 ± 0.11	0.62 ± 0.06
Breast-w	0.96 ± 0.01	0.95 ± 0.02	0.95 ± 0.02	0.94 ± 0.03
Bridges2	0.70 ± 0.02	0.43 ± 0.05	0.51 ± 0.11	0.29 ± 0.11
Car	0.92 ± 0.03	0.97 ± 0.02	0.93 ± 0.02	0.89 ± 0.02
Heart-c	0.83 ± 0.01	0.80 ± 0.04	0.76 ± 0.09	0.77 ± 0.08
Flags	0.72 ± 0.01	0.57 ± 0.10	0.61 ± 0.05	0.58 ± 0.10
Credit-g	0.76 ± 0.01	0.71 ± 0.04	0.70 ± 0.02	0.70 ± 0.05
Colic	0.88 ± 0.01	0.84 ± 0.08	0.85 ± 0.07	0.83 ± 0.07
Heart-h	0.83 ± 0.01	0.76 ± 0.06	0.80 ± 0.07	0.79 ± 0.09
Ionosphere	0.94 ± 0.01	0.89 ± 0.03	0.88 ± 0.05	0.91 ± 0.02
kdd-synthetic	0.95 ± 0.00	0.88 ± 0.04	0.91 ± 0.04	0.87 ± 0.04
kr-vs-kp	0.96 ± 0.00	0.99 ± 0.01	0.99 ± 0.01	0.99 ± 0.01
Liver-disorders	0.75 ± 0.00	0.66 ± 0.09	0.66 ± 0.05	0.63 ± 0.05
Lung-cancer	0.71 ± 0.02	0.42 ± 0.32	0.35 ± 0.29	0.42 ± 0.19
Meta.data	0.13 ± 0.02	0.02 ± 0.01	0.02 ± 0.02	0.00 ± 0.00
Morphological	0.72 ± 0.00	0.70 ± 0.04	0.70 ± 0.02	0.70 ± 0.03
mb-promoters	0.89 ± 0.01	0.71 ± 0.14	0.79 ± 0.14	0.76 ± 0.15
Shuttle-control	0.55 ± 0.02	0.57 ± 0.39	0.57 ± 0.39	0.57 ± 0.39
Soybean	0.80 ± 0.09	0.91 ± 0.04	0.92 ± 0.04	0.82 ± 0.06
Tempdiag	1.00 ± 0.00	1.00 ± 0.00	1.00 ± 0.00	1.00 ± 0.00
Tep.fea	0.61 ± 0.00	0.61 ± 0.02	0.61 ± 0.02	0.61 ± 0.02
Tic-tac-toe	0.94 ± 0.02	0.94 ± 0.02	0.86 ± 0.03	0.86 ± 0.03
Transfusion	0.77 ± 0.00	0.76 ± 0.03	0.77 ± 0.03	0.76 ± 0.02
Vehicle	0.77 ± 0.01	0.71 ± 0.05	0.73 ± 0.04	0.70 ± 0.04
Vote	0.96 ± 0.00	0.97 ± 0.02	0.97 ± 0.02	0.95 ± 0.03
Vowel	0.76 ± 0.07	0.82 ± 0.04	0.83 ± 0.03	0.70 ± 0.04
Wine-red	0.63 ± 0.01	0.61 ± 0.02	0.61 ± 0.03	0.58 ± 0.03
Wine-white	0.53 ± 0.01	0.58 ± 0.03	0.60 ± 0.02	0.55 ± 0.02
Average rank	**1.71**	2.53	2.29	3.47

Table 5.21 Results for the $\{9 \times 31\}$ configuration

(a) *Accuracy results*

Data set	HEAD	CART	C4.5	REP
Abalone	0.28 ± 0.00	0.26 ± 0.02	0.22 ± 0.02	0.26 ± 0.02
Anneal	0.92 ± 0.00	0.98 ± 0.01	0.99 ± 0.01	0.98 ± 0.02
Arrhythmia	0.76 ± 0.00	0.71 ± 0.05	0.66 ± 0.05	0.67 ± 0.06
Audiology	0.67 ± 0.01	0.74 ± 0.05	0.78 ± 0.07	0.74 ± 0.08
Autos	0.78 ± 0.01	0.78 ± 0.10	0.86 ± 0.06	0.65 ± 0.08
Breast-cancer	0.73 ± 0.00	0.69 ± 0.04	0.75 ± 0.08	0.69 ± 0.05
Breast-w	0.94 ± 0.00	0.95 ± 0.02	0.95 ± 0.02	0.94 ± 0.03
Bridges2	0.62 ± 0.01	0.54 ± 0.08	0.58 ± 0.13	0.40 ± 0.15
Car	0.88 ± 0.01	0.97 ± 0.02	0.93 ± 0.02	0.89 ± 0.02
Heart-c	0.82 ± 0.00	0.81 ± 0.04	0.77 ± 0.09	0.77 ± 0.08
Flags	0.71 ± 0.01	0.61 ± 0.10	0.63 ± 0.05	0.62 ± 0.10
Credit-g	0.75 ± 0.00	0.73 ± 0.04	0.71 ± 0.03	0.72 ± 0.06
Colic	0.83 ± 0.01	0.85 ± 0.08	0.86 ± 0.06	0.83 ± 0.06
Heart-h	0.80 ± 0.00	0.77 ± 0.06	0.80 ± 0.08	0.80 ± 0.09
Ionosphere	0.92 ± 0.00	0.89 ± 0.03	0.89 ± 0.05	0.91 ± 0.02
kdd-synthetic	0.95 ± 0.00	0.88 ± 0.04	0.91 ± 0.04	0.88 ± 0.03
kr-vs-kp	0.91 ± 0.01	0.99 ± 0.01	0.99 ± 0.01	0.99 ± 0.01
Liver-disorders	0.74 ± 0.00	0.67 ± 0.09	0.67 ± 0.05	0.65 ± 0.05
Meta.data	0.11 ± 0.00	0.05 ± 0.03	0.04 ± 0.03	0.04 ± 0.00
Morphological	0.71 ± 0.00	0.72 ± 0.04	0.72 ± 0.02	0.72 ± 0.03
mb-promoters	0.73 ± 0.00	0.72 ± 0.14	0.80 ± 0.13	0.77 ± 0.15
Shuttle-control	0.63 ± 0.00	0.65 ± 0.34	0.65 ± 0.34	0.65 ± 0.34
Soybean	0.64 ± 0.01	0.92 ± 0.04	0.92 ± 0.03	0.84 ± 0.05
Tep.fea	0.65 ± 0.00	0.65 ± 0.02	0.65 ± 0.02	0.65 ± 0.02
Tic-tac-toe	0.83 ± 0.01	0.94 ± 0.02	0.86 ± 0.03	0.86 ± 0.03
Transfusion	0.79 ± 0.00	0.79 ± 0.03	0.78 ± 0.02	0.78 ± 0.02
Vehicle	0.75 ± 0.00	0.72 ± 0.04	0.74 ± 0.04	0.71 ± 0.04
Vote	0.96 ± 0.00	0.97 ± 0.02	0.97 ± 0.02	0.95 ± 0.03
Vowel	0.61 ± 0.01	0.82 ± 0.04	0.83 ± 0.03	0.70 ± 0.04
Wine-red	0.61 ± 0.00	0.63 ± 0.02	0.61 ± 0.03	0.60 ± 0.03
Wine-white	0.53 ± 0.00	0.58 ± 0.02	0.61 ± 0.03	0.56 ± 0.02
Average rank	2.55	2.26	**2.03**	3.16

(b) *F-Measure results*

Data set	HEAD	CART	C4.5	REP
Abalone	0.24 ± 0.00	0.23 ± 0.02	0.21 ± 0.02	0.24 ± 0.02
Anneal	0.92 ± 0.00	0.98 ± 0.01	0.98 ± 0.01	0.98 ± 0.02
Arrhythmia	0.73 ± 0.00	0.67 ± 0.06	0.65 ± 0.06	0.63 ± 0.07
Audiology	0.62 ± 0.01	0.71 ± 0.05	0.75 ± 0.08	0.70 ± 0.09

(continued)

Table 5.21 (continued)

(b) *F-Measure results*

Data set	HEAD	CART	C4.5	REP
Autos	0.78 ± 0.01	0.77 ± 0.10	0.85 ± 0.07	0.62 ± 0.07
Breast-cancer	0.71 ± 0.00	0.63 ± 0.05	0.70 ± 0.11	0.62 ± 0.06
Breast-w	0.94 ± 0.00	0.95 ± 0.02	0.95 ± 0.02	0.94 ± 0.03
Bridges2	0.60 ± 0.01	0.43 ± 0.05	0.51 ± 0.11	0.29 ± 0.11
Car	0.88 ± 0.01	0.97 ± 0.02	0.93 ± 0.02	0.89 ± 0.02
Heart-c	0.82 ± 0.00	0.80 ± 0.04	0.76 ± 0.09	0.77 ± 0.08
Flags	0.70 ± 0.00	0.57 ± 0.10	0.61 ± 0.05	0.58 ± 0.10
Credit-g	0.73 ± 0.00	0.71 ± 0.04	0.70 ± 0.02	0.70 ± 0.05
Colic	0.83 ± 0.01	0.84 ± 0.08	0.85 ± 0.07	0.83 ± 0.07
Heart-h	0.79 ± 0.00	0.76 ± 0.06	0.80 ± 0.07	0.79 ± 0.09
Ionosphere	0.92 ± 0.00	0.89 ± 0.03	0.88 ± 0.05	0.91 ± 0.02
kdd-synthetic	0.95 ± 0.00	0.88 ± 0.04	0.91 ± 0.04	0.87 ± 0.04
kr-vs-kp	0.91 ± 0.01	0.99 ± 0.01	0.99 ± 0.01	0.99 ± 0.01
Liver-disorders	0.73 ± 0.00	0.66 ± 0.09	0.66 ± 0.05	0.63 ± 0.05
Meta.data	0.09 ± 0.00	0.02 ± 0.01	0.02 ± 0.02	0.00 ± 0.00
Morphological	0.69 ± 0.00	0.70 ± 0.04	0.70 ± 0.02	0.70 ± 0.03
mb-promoters	0.73 ± 0.00	0.71 ± 0.14	0.79 ± 0.14	0.76 ± 0.15
Shuttle-control	0.56 ± 0.00	0.57 ± 0.39	0.57 ± 0.39	0.57 ± 0.39
Soybean	0.60 ± 0.02	0.91 ± 0.04	0.92 ± 0.04	0.82 ± 0.06
Tep.fea	0.61 ± 0.00	0.61 ± 0.02	0.61 ± 0.02	0.61 ± 0.02
Tic-tac-toe	0.83 ± 0.02	0.94 ± 0.02	0.86 ± 0.03	0.86 ± 0.03
Transfusion	0.77 ± 0.00	0.76 ± 0.03	0.77 ± 0.03	0.76 ± 0.02
Vehicle	0.74 ± 0.00	0.71 ± 0.05	0.73 ± 0.04	0.70 ± 0.04
Vote	0.96 ± 0.00	0.97 ± 0.02	0.97 ± 0.02	0.95 ± 0.03
Vowel	0.60 ± 0.01	0.82 ± 0.04	0.83 ± 0.03	0.70 ± 0.04
Wine-red	0.59 ± 0.00	0.61 ± 0.02	0.61 ± 0.03	0.58 ± 0.03
Wine-white	0.49 ± 0.00	0.58 ± 0.03	0.60 ± 0.02	0.55 ± 0.02
Average rank	2.48	2.32	**1.96**	3.23

The last step of this empirical analysis is to verify whether the differences in rank values are statistically significant. We employ once again the graphical representation suggested by Demšar [7], the *critical diagrams*. Algorithms that are not significantly different from each other are connected. The critical difference given by the Nemenyi test is presented in the top of the graph.

Figure 5.3 shows the critical diagrams for all experimental configurations. It is interesting to see that there are no statistically significant differences among HEAD-DT, C4.5, and CART in neither configuration, except for $\{7 \times 33\}$, in which HEAD-DT outperforms CART with statistical significance (see Fig. 5.3h). Indeed, this is the only case in which one of the three algorithms outperform another in all five

Table 5.22 Summary of the experimental analysis regarding the heterogeneous approach

Configuration	Rank	HEAD-DT	CART	C4.5	REPTree
{1 × 39}	Accuracy	**2.24**	2.36	**2.24**	3.15
	F-measure	2.18	2.46	**2.15**	3.21
{3 × 37}	Accuracy	2.45	2.26	**2.19**	3.11
	F-measure	2.23	2.42	2.15	3.20
{5 × 35}	Accuracy	2.56	2.19	**2.13**	3.13
	F-measure	2.44	2.27	**2.04**	3.24
{7 × 33}	Accuracy	**1.77**	2.50	2.35	3.38
	F-measure	**1.71**	2.53	2.29	3.47
{9 × 31}	Accuracy	2.55	2.26	**2.03**	3.16
	F-measure	2.48	2.32	**1.96**	3.23
	Average	2.26	2.36	**2.15**	3.23

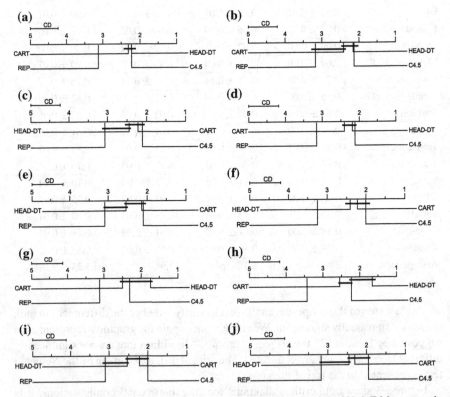

Fig. 5.3 Critical diagrams for the UCI data sets. **a** Accuracy rank for {1 × 39}. **b** F-Measure rank for {1 × 39}. **c** Accuracy rank for {3 × 37}. **d** F-Measure rank for {3 × 37}. **e** Accuracy rank for {5 × 35}. **f** F-Measure rank for {5 × 35}. **g** Accuracy rank for {7 × 33}. **h** F-Measure rank for {7 × 33}. **i** Accuracy rank for {9 × 31}. **j** F-Measure rank for {9 × 31}

configurations. REPTree is often outperformed with statistical significance by the three other methods, which is expected considering it is consistently the worst-ranked algorithm among the four.

Note that HEAD-DT did not achieve the same performance as in the homogeneous approach. That seems to show that generating an "all-around" algorithm is more difficult than generating a tailor-made algorithm for a particular domain. This is due to the own nature of the optimisation strategy employed by HEAD-DT. By optimising the building blocks of a decision-tree induction algorithm based on a few data sets in the meta-training set, HEAD-DT eventually finds a good compromise regarding the average F-Measure of these data sets. Nevertheless, we believe this automatically-designed algorithm to be too-specific to the meta-training data sets, leading to a specific case of overfitting. Our hypothesis is that HEAD-DT's generated algorithm is overfitting (at some extent) the meta-training set, causing some damage to its performance over the meta-test set. We further investigate this hypothesis in the next section, backing it up with data.

5.2.3 The Case of Meta-Overfitting

When evolving an algorithm from multiple data sets, HEAD-DT optimises its algorithms based on their predictive performance on a few data sets in the meta-training set. We saw in the previous section that generating an "all-around" algorithm that is capable of outperforming traditional algorithms, such as C4.5 and CART, is not an easy task. Our hypothesis for this apparent "lack of performance" is that HEAD-DT is finding a good (perhaps excellent) algorithm for the meta-training set, but that this algorithm is not really a good "all-around" algorithm. We call this phenomenon *meta-overfitting*.[4]

For supporting this hypothesis with data, let us analyse the performance of HEAD-DT in the meta-training set considering the previously-detailed configurations. Table 5.23 shows the F-Measure obtained by HEAD-DT in each data set in the meta-training set, as well as their average (fitness value). For a perspective view, we also present the same results for C4.5, considering the exact same data partition. Recall that, during the fitness evaluation process, we employ a random training-test partition for each data set, instead of typical 10-fold cross-validation procedure, in order to speed up evolution.

Table 5.23 shows that HEAD-DT is indeed designing algorithms that excel in the meta-training set. The average F-Measure achieved by HEAD-DT in the meta-training set is always greater than the one provided by C4.5. Whereas it is expected that HEAD-DT generates algorithms that perform well in the meta-training set (it is explicitly optimising these algorithms for that goal), the difference in perfor-

---

[4] The term *overfitting* is not used because it refers to a model that overfits the data, whereas we are talking about the case of an *algorithm* that "overfits" the data, in the sense that it is excellent when dealing with those data sets it was designed to, but it underperforms in previously unseen data sets.

Table 5.23 F-Measure achieved by HEAD-DT and C4.5 in the meta-training set

Configuration	Data sets	HEAD-DT	C4.5	Average HEAD-DT	Average C4.5
{1 × 39}	Hayes-roth	**0.9242**	0.7708	**0.9242**	0.7708
{3 × 37}	Hayes-roth	**0.9242**	0.7708	**0.8438**	0.7352
	Labor	**1.0000**	0.8667		
	Tae	**0.6073**	0.5682		
{5 × 35}	Hayes-roth	**0.9242**	0.7708	**0.8494**	0.7565
	Labor	**1.0000**	0.8667		
	Tae	**0.6073**	0.5682		
	Iris	0.9471	0.9471		
	Haberman	**0.7683**	0.6298		
{7 × 33}	Hayes-roth	**0.9242**	0.7708	**0.7808**	0.6940
	Labor	0.8667	0.8667		
	Tae	**0.6017**	0.5682		
	Iris	**0.9736**	0.9471		
	Haberman	**0.7697**	0.6298		
	Postoperative-patient	**0.6293**	0.5708		
	Bridges1	**0.7003**	0.5049		
{9 × 31}	Hayes-roth	**0.8750**	0.7708	**0.7687**	0.6628
	Labor	0.8038	**0.8667**		
	Tae	**0.6223**	0.5682		
	Iris	0.9471	0.9471		
	haberman	**0.7342**	0.6298		
	Postoperative-patient	0.5708	0.5708		
	Bridges1	**0.6190**	0.5049		
	Tempdiag	1.0000	1.0000		
	Lung-cancer	**0.7464**	0.1071		

mance between HEAD-DT and C4.5 is overwhelming. For instance, in configurations {3 × 37}, {5 × 35}, {7 × 33}, and {9 × 31}, HEAD-DT provides F-Measure values that are around 10 % higher than those provided by C4.5. In configuration {1 × 39}, in which HEAD-DT has to optimise a decision-tree induction algorithm based on the sole predictive performance of data set *hayes-roth*, the difference in F-Measure is even higher: 15 %!

These very good results achieved by HEAD-DT in the meta-training set and, at the same time, its disappointing results in the corresponding meta-test sets (except for configuration {7 × 33}) seem to indicate that HEAD-DT is suffering from *meta-overfitting*. Unfortunately, the problem does not have a trivial solution. We comment on possible solutions to this problem in the book's future work possibilities, Sect. 7.2.

5.3 HEAD-DT's Time Complexity

Regarding execution time, it is clear that HEAD-DT is slower than C4.5, CART, or REPTree. Considering that there are 100 individuals executed for 100 generations, there is a maximum (worst case) of 10,000 fitness evaluations of decision trees.

We recorded the execution time of both breeding operations and fitness evaluation (one thread was used for breeding and other for evaluation). Total time of breeding is negligible (a few milliseconds in a full evolutionary cycle), regardless of the data sets being used in the meta-training set (breeding does not consider any domain-specific information). Indeed, breeding individuals in the form of an integer string is known to be quite efficient in evolutionary computation.

Fitness evaluation, on the other hand, is the bottleneck of HEAD-DT. In the experiments of the specific framework, the largest data set (winequality_white) takes 2.5 h to be fully executed (one full evolutionary cycle of 100 generations). The smallest data set (shuttle_landing) takes only 0.72 s to be fully executed. In the homogeneous approach of the general framework, the most time-consuming configuration, $\{9 \times 12\}$, takes 11.62 h to be fully executed, whereas the fastest configuration, $\{1 \times 20\}$, takes only 5.60 min. Hence, we can see that the fitness evaluation time can vary quite a lot based on the number and type of data sets in the meta-training set.

The computational complexity of top-down decision-tree induction algorithms like C4.5 and CART is $O(m \times n \log n)$ (m is the number of attributes and n the data set size), plus a term regarding the specific pruning method. Considering that breeding takes negligible time, we can say that in the worst case scenario, HEAD-DT time complexity is $O(i \times g \times m \times n \log n)$, where i is the number of individuals and g is the number of generations. In practice, the number of evaluations is much smaller than $i \times g$, due to the fact that repeated individuals are not re-evaluated. In addition, individuals selected by elitism and by reproduction (instead of crossover) are also not re-evaluated, saving computational time.

5.4 Cost-Effectiveness of Automated Versus Manual Algorithm Design

The aforementioned conventional perspective for analysing HEAD-DT's time complexity is misleading in one way: it assumes that HEAD-DT is a "conventional" search algorithm, searching for an optimal solution to a single data set (as usual when running a classification algorithm), which is not the case. In reality, as discussed earlier, HEAD-DT is a hyper-heuristic that outputs a complete decision-tree induction algorithm. The algorithm automatically designed by HEAD-DT, as well as the manually-designed algorithms C4.5, CART and REPTree, are all complete decision-tree induction algorithms that can be re-used over and over again to extract knowledge from different data sets. In the machine learning literature, the time taken by human researchers to manually design an algorithm is usually not reported, but it is safe to assume that the time taken by a single human researcher to design and implement a new decision-tree induction algorithm is on the order of at least several months. In this context, noting that what HEAD-DT is doing is effectively replacing the manual design of decision-tree algorithms with an automated approach for such a design, even if HEAD-DT took a couple of days to produce a decision-tree algorithm, that time would still be much smaller than the corresponding manual design time. Hence, when evaluated as an algorithm-design method (which is really the role of HEAD-DT), it is fair to say it is a very fast method, at least much faster than a manual approach for the design of decision-tree induction algorithms.

At this point in our discussion, trying to play the role of devil's lawyer, one could perhaps present the following counter-argument: the aforementioned discussion is ignoring the fact that HEAD-DT was itself designed by human researchers, a design process that also took several months! This is of course true, but even taking this into account, it can be argued that HEAD-DT is still much more cost-effective than the human design of decision-tree algorithms, as follows. First, now that HEAD-DT has been manually designed, it can be re-used over and over again to automatically create decision-tree algorithms tailored to any particular type of data set (or application domain) in which a given user is interested. In the general framework, we focused on gene expression data sets, but HEAD-DT can be re-used to create decision-tree algorithms tailored to, say, a specific type of financial data sets or a specific type of medical data sets, to mention just two out of a very large number of possible application domains. Once HEAD-DT has been created, the "cost" associated with using HEAD-DT in any other application domain is very small—it is essentially the cost associated with the time to run HEAD-DT in a new application domain (say a couple of days of a desktop computer's processing time).

In contrast, what would be the cost of manually creating a new decision-tree algorithm tailored to a particular type of data set or application domain? First, note that this kind of manual design of a decision-tree algorithm tailored to a specific type of data set is hardly found in the literature. This is presumably because, for the new algorithm to be effective and really tailored to the target type of data set, a human researcher would need to be an expert in both decision-tree algorithms and the application domain (or more specifically the type of data set to be mined), and not many researchers would satisfy both criteria in practice. Just for the sake of argument, though, let us make the (probably unrealistic) assumption that there are many application domains for which there is a researcher who is an expert in both that application domain and decision-tree induction algorithms. For each such application domain, it seems safe to assume again that the human expert in question would need on the order of several months to design a new decision-tree algorithm that is effective and really tailored to that application domain; whilst, as mentioned earlier, HEAD-DT could automatically design this algorithm in a couple of days.

In summary, given the very large diversity of application domains to which decision-tree algorithms have been and will probably continue to be applied for a long time, one can see that HEAD-DT's automated approach offers a much more cost-effective approach for designing decision-tree algorithms than the conventional manual design approach that is nearly always used in machine learning research. In this sense, HEAD-DT paves the way for the large-scale and cost-effective production of decision-tree induction algorithms that are tailored to any specific application domain or type of classification data set of interest.

5.5 Examples of Automatically-Designed Algorithms

For illustrating a novel algorithm designed by HEAD-DT, let us first consider the specific framework, more specifically the algorithm designed to the Semeion data set, in which HEAD-DT managed to achieve maximum accuracy and F-Measure (which was not the case of CART and C4.5). The algorithm designed by HEAD-DT is presented in Algorithm 1. It is indeed novel, since no algorithm in the literature combines components such as the Chandra-Varghese criterion with a pruning-free strategy.

Algorithm 1 Algorithm designed by HEAD-DT for the Semeion data set.

1: Recursively split nodes with the Chandra-Varghese criterion;
2: Aggregate nominal splits in binary subsets;
3: Perform step 1 until class-homogeneity or the minimum number of 5 instances is reached;
4: Do not perform any pruning;
   When dealing with missing values:
5: Calculate the split of missing values by performing unsupervised imputation;
6: Distribute missing values by assigning the instance to all partitions;
7: For classifying an instance with missing values, explore all branches and combine the results.

The main advantage of HEAD-DT is that it automatically searches for the suitable components (with their own biases) according to the data set (or set of data sets) being investigated. It is difficult to believe that a researcher would combine such a distinct set of components like those in Algorithm 1 to achieve 100 % accuracy in a particular data set.

Now let us consider the general framework, more specifically the homogeneous approach, in which HEAD-DT managed to outperform C4.5, CART, and REPTree for both accuracy and F-Measure. The typical algorithm designed by HEAD-DT for the domain of gene expression classification is presented in Algorithm 2. It is indeed novel, since no algorithm in the literature combines components such as the DCSM criterion with PEP pruning.

The algorithm presented in Algorithm 2 is one of the twenty-five algorithms automatically designed by HEAD-DT in the experimental analysis (5 configurations executed 5 times each). Nevertheless, by close inspection of the 25 automatically-generated algorithms, we observed that Algorithm 2 comprises building blocks that were consistently favored regarding the gene expression application domain. For instance, the DCSM and Chandra-Varghese criteria—both created recently by the same authors [5, 6]—were selected as the best split criterion in ≈50 % of the algorithms designed by HEAD-DT. Similarly, the *minimum number of instances* stop criterion was selected in ≈70 % of the algorithms, with either 6 or 7 instances as its

---

Algorithm 2 Algorithm designed by HEAD-DT for the homogeneous approach (gene expression data), configuration $\{5 \times 16\}$.

1: Recursively split nodes using the DCSM criterion;
2: Aggregate nominal splits in binary subsets;
3: Perform step 1 until class-homogeneity or the minimum number of 6 instances is reached;
4: Perform PEP pruning with 2 standard errors (SEs) to adjust the training error;
   When dealing with missing values:
5: Calculate the split of missing values by performing unsupervised imputation;
6: Distribute missing values by assigning the instance to the largest partition;
7: For classifying an instance with missing values, explore all branches and combine the results.

---

parameter value. Finally, the PEP pruning was the favored pruning strategy in $\approx 60\%$ of the algorithms, with a very large advantage over the strategies used by C4.5 (EBP pruning, selected in 8 % of the algorithms) and CART (CCP pruning, not selected by any of the automatically-designed algorithms).

5.6 Is the Genetic Search Worthwhile?

Finally, the last task to be performed in this chapter is to verify whether the genetic search employed by HEAD-DT provides solutions statistically better than a random walk through the search space. For that, we implemented the random search algorithm depicted in Algorithm 3.

The random search algorithm searches in the space of 10,000 individuals in order to make a fair comparison with an evolutionary algorithm that evolves 100 individuals in 100 generations. These 10,000 individuals are randomly created by generating random values from a uniform distribution to each of the individual's genes, within its valid boundaries (Algorithm 3, line 5).

After the genome is decoded in the form of a decision-tree induction algorithm (Algorithm 3, line 7), the randomly-generated algorithm is executed over the meta-training set (Algorithm 3, lines 9–11), and its fitness is computed as the average F-Measure achieved in each data set from the meta-training set (Algorithm 3, line 12). The best individual is stored in the variable *bestAlgorithm* (Algorithm 9, lines 13–16), and returned as the resulting decision-tree algorithm from the random search procedure.

In order to compare the algorithms generated by HEAD-DT with those created by the random search algorithm, we employed the same experimental setup presented in Sect. 5.2.2: UCI data sets divided in 5 configurations: {#meta-training sets, #meta-test sets}: $\{1 \times 39\}$, $\{3 \times 37\}$, $\{5 \times 35\}$, $\{7 \times 33\}$, and $\{9 \times 31\}$. For evaluating the statistical significance of the results, we applied the Wilcoxon signed-ranks test [13], which is the recommended statistical test to evaluate two classifiers in multiple data sets [7]. In a nutshell, the Wilcoxon signed-ranks test is a non-parametric alternative to the well-known paired t-test, which ranks the differences in performances of two classifiers for each data set (ignoring their signs) and compares the ranks for the positive and negative differences. The sum of the positive and negative ranks are

Algorithm 3 Random search algorithm.

```
 1: bestFitness ← 0
 2: bestAlgorithm ← ∅
 3: for i = 1 to 10000 do
 4:     for gene in genome do
 5:         gene ← U(min_gene, max_gene)
 6:     end for
 7:     algorithm ← decode(genome)
 8:     for dataset in meta-training do
 9:         DT ← execute(algorithm, dataset.training)
10:         fmeasure[dataset] ← classify(DT, dataset.test)
11:     end for
12:     fitness[i] ← average(fmeasure)
13:     if (fitness[i] > bestFitness) then
14:         bestFitness ← fitness[i]
15:         bestAlgorithm ← algorithm
16:     end if
17: end for
18: return bestAlgorithm
```

performed, and the smaller of them is compared to a table of exact critical values (for up to 25 data sets), and for larger values a z statistic is computed and assumed to be normally distributed.

Tables 5.24, 5.25, 5.26, 5.27 and 5.28 present the comparison results from the 5 configurations, for both accuracy and F-Measure. At the bottom of each table, we present the number of victories for each method (ties are omitted), and also the p-value returned by the Wilcoxon test. For rejecting the null hypothesis of performance equivalency between the two algorithms, the p-values should be smaller than the desired significance level α. Note that, regardless of the configuration, the p-values are really small, showing that HEAD-DT is significantly better than the random search algorithm considering $\alpha = 0.05$ and also $\alpha = 0.01$. These results support the hypothesis that the genetic search employed by HEAD-DT is indeed worthwhile, allowing an effective search in the space of decision-tree induction algorithms.

5.7 Chapter Remarks

In this chapter, we presented two distinct sets of experiments for assessing the effectiveness of HEAD-DT, according to the fitness evaluation frameworks presented in Chap. 4. In the first set of experiments, which concerned the specific framework, both the meta-training and meta-test sets comprise data belonging to a single data set. We evaluated the performance of algorithms automatically designed by HEAD-DT in 20 public UCI data sets, and we compared their performance with C4.5 and CART. Results showed that HEAD-DT is capable of generating specialized algorithms whose performance is significantly better than that of the baseline algorithms.

In the second set of experiments, which concerned the general framework, we evaluated HEAD-DT in two different scenarios: (i) the homogeneous approach, in which HEAD-DT evolved a single algorithm to be applied in data sets from a particular

Table 5.24   HEAD-DT versus random search: $\{1 \times 39\}$ configuration

(a) *Accuracy results*

Data set	HEAD-DT	Random search
Abalone	0.36 ± 0.10	0.25 ± 0.01
Anneal	0.92 ± 0.05	0.93 ± 0.10
Arrhythmia	0.66 ± 0.08	0.58 ± 0.04
Audiology	0.70 ± 0.06	0.56 ± 0.19
Autos	0.80 ± 0.01	0.64 ± 0.16
Breast-cancer	0.73 ± 0.02	0.68 ± 0.04
Breast-w	0.94 ± 0.02	0.94 ± 0.01
Bridges1	0.70 ± 0.03	0.56 ± 0.11
Bridges2	0.64 ± 0.07	0.56 ± 0.11
Car	0.86 ± 0.07	0.88 ± 0.04
Heart-c	0.83 ± 0.02	0.76 ± 0.03
Flags	0.71 ± 0.04	0.56 ± 0.15
Credit-g	0.78 ± 0.03	0.72 ± 0.01
Colic	0.74 ± 0.07	0.70 ± 0.10
Haberman	0.77 ± 0.01	0.72 ± 0.02
Heart-h	0.83 ± 0.02	0.76 ± 0.07
Ionosphere	0.90 ± 0.03	0.89 ± 0.02
Iris	0.96 ± 0.01	0.95 ± 0.01
kdd-synthetic	0.93 ± 0.03	0.88 ± 0.03
kr-vs-kp	0.91 ± 0.06	0.87 ± 0.20
Labor	0.79 ± 0.06	0.69 ± 0.04
Liver-disorders	0.77 ± 0.04	0.67 ± 0.02
Lung-cancer	0.66 ± 0.03	0.45 ± 0.06
Meta.data	0.11 ± 0.02	0.03 ± 0.01
Morphological	0.73 ± 0.03	0.70 ± 0.02
mb-promoters	0.78 ± 0.07	0.76 ± 0.03
Postoperative-patient	0.69 ± 0.01	0.67 ± 0.05
Shuttle-control	0.57 ± 0.04	0.51 ± 0.10
Soybean	0.69 ± 0.20	0.66 ± 0.30
Tae	0.64 ± 0.03	0.51 ± 0.06
Tempdiag	0.97 ± 0.04	0.97 ± 0.05
Tep.fea	0.65 ± 0.00	0.65 ± 0.00
Tic-tac-toe	0.83 ± 0.08	0.83 ± 0.05
Transfusion	0.80 ± 0.01	0.77 ± 0.02
Vehicle	0.78 ± 0.04	0.70 ± 0.03
Vote	0.95 ± 0.00	0.95 ± 0.01
Vowel	0.72 ± 0.16	0.63 ± 0.09

(continued)

Table 5.24   (continued)

(a) *Accuracy results*

Data set	HEAD-DT	Random search
Wine-red	0.67 ± 0.06	0.57 ± 0.01
Wine-white	0.62 ± 0.10	0.53 ± 0.02
Number of victories	34	4
Wilcoxon p-value	1.62×10^{-9}	

(b) *F-Measure results*

Data set	HEAD-DT	Random search
Abalone	0.34 ± 0.11	0.77 ± 0.01
Anneal	0.90 ± 0.08	0.09 ± 0.14
Arrhythmia	0.60 ± 0.11	0.53 ± 0.10
Audiology	0.66 ± 0.06	0.50 ± 0.22
Autos	0.80 ± 0.01	0.39 ± 0.19
Breast-cancer	0.71 ± 0.03	0.36 ± 0.05
Breast-w	0.94 ± 0.02	0.06 ± 0.01
Bridges1	0.68 ± 0.04	0.50 ± 0.13
Bridges2	0.62 ± 0.08	0.51 ± 0.12
Car	0.85 ± 0.08	0.12 ± 0.04
Heart-c	0.83 ± 0.02	0.24 ± 0.04
Flags	0.70 ± 0.05	0.48 ± 0.21
Credit-g	0.77 ± 0.04	0.32 ± 0.06
Colic	0.72 ± 0.09	0.37 ± 0.15
Haberman	0.75 ± 0.01	0.31 ± 0.02
Heart-h	0.83 ± 0.02	0.27 ± 0.11
Ionosphere	0.90 ± 0.04	0.12 ± 0.02
Iris	0.96 ± 0.01	0.05 ± 0.01
kdd-synthetic	0.93 ± 0.03	0.12 ± 0.03
kr-vs-kp	0.91 ± 0.06	0.16 ± 0.27
Labor	0.76 ± 0.09	0.38 ± 0.05
Liver-disorders	0.77 ± 0.04	0.33 ± 0.02
Lung-cancer	0.65 ± 0.04	0.64 ± 0.10
Meta.data	0.10 ± 0.03	0.98 ± 0.01
Morphological	0.71 ± 0.04	0.31 ± 0.01
mb-promoters	0.78 ± 0.07	0.24 ± 0.03
Postoperative-patient	0.64 ± 0.02	0.42 ± 0.02
Shuttle-control	0.55 ± 0.06	0.60 ± 0.05
Soybean	0.66 ± 0.21	0.37 ± 0.33
Tae	0.64 ± 0.03	0.49 ± 0.06
Tempdiag	0.97 ± 0.04	0.03 ± 0.05
Tep.fea	0.61 ± 0.00	0.39 ± 0.00

(continued)

Table 5.24 (continued)

(b) *F-Measure results*

Data set	HEAD-DT	Random search
Tic-tac-toe	0.83 ± 0.09	0.18 ± 0.05
Transfusion	0.78 ± 0.01	0.25 ± 0.01
Vehicle	0.77 ± 0.04	0.30 ± 0.03
Vote	0.95 ± 0.00	0.05 ± 0.01
Vowel	0.71 ± 0.17	0.37 ± 0.09
Wine-red	0.66 ± 0.07	0.44 ± 0.02
Wine-white	0.60 ± 0.12	0.49 ± 0.03
Number of victories	36	3
Wilcoxon p-value	2.91×10^{-7}	

application domain, namely microarray gene expression data classification; and (ii) the heterogeneous approach, in which HEAD-DT evolved a single algorithm to be applied in a variety of data sets, aiming at generating an effective "all-around" algorithm.

In the homogeneous approach, HEAD-DT was the best-ranked method in all configurations, supporting the hypothesis that HEAD-DT is indeed capable of generating an algorithm tailor-made to a particular application domain. In the heterogeneous approach, HEAD-DT presented predictive performance similar to C4.5 and CART— no statistically-significant difference was found, except for a particular configuration in which HEAD-DT outperformed CART. By further analysing the data collected from the meta-training set, we concluded that HEAD-DT may be suffering from *meta-overfitting*, generating algorithms that excel in the meta-training set, but that underperform at previously-unseen data sets. Considering that CART and C4.5 algorithms are effective and efficient "all-around" algorithms, we recommend their use in those scenarios in which the user needs fast results in a broad and unrelated set of data.

We believe the sets of experiments presented in this chapter were comprehensive enough to conclude that HEAD-DT is an effective approach for building specialized decision-tree induction algorithms tailored to particular domains or individual data sets. We refer the interested reader to two successful domain-based applications of HEAD-DT: flexible-receptor molecular docking data [1] and software maintenance effort prediction [3].

We also presented in this chapter the time complexity of HEAD-DT, as well as two examples of automatically-designed decision-tree algorithms. Finally, we investigated whether the genetic search employed by HEAD-DT was really worthwhile by comparing it to a random-search strategy. Results clearly indicated that the evolutionary process effectively guides the search for robust decision-tree algorithms.

Table 5.25 HEAD-DT versus random search: $\{3 \times 37\}$ configuration

(a) *Accuracy results*

Data set	HEAD-DT	Random search
Abalone	0.27 ± 0.00	0.25 ± 0.01
Anneal	0.98 ± 0.01	0.93 ± 0.10
Arrhythmia	0.66 ± 0.10	0.58 ± 0.04
Audiology	0.73 ± 0.02	0.56 ± 0.19
Autos	0.73 ± 0.08	0.64 ± 0.16
Breast-cancer	0.74 ± 0.01	0.68 ± 0.04
Breast-w	0.94 ± 0.01	0.94 ± 0.01
Bridges1	0.70 ± 0.06	0.56 ± 0.11
Bridges2	0.69 ± 0.05	0.56 ± 0.11
Car	0.84 ± 0.02	0.88 ± 0.04
Heart-c	0.81 ± 0.00	0.76 ± 0.03
Flags	0.69 ± 0.01	0.56 ± 0.15
Credit-g	0.74 ± 0.00	0.72 ± 0.01
Colic	0.78 ± 0.12	0.70 ± 0.10
Haberman	0.77 ± 0.00	0.72 ± 0.02
Heart-h	0.80 ± 0.01	0.76 ± 0.07
Ionosphere	0.92 ± 0.03	0.89 ± 0.02
Iris	0.96 ± 0.00	0.95 ± 0.01
kdd-synthetic	0.95 ± 0.01	0.88 ± 0.03
kr-vs-kp	0.91 ± 0.03	0.87 ± 0.20
Liver-disorders	0.73 ± 0.01	0.67 ± 0.02
Lung-cancer	0.69 ± 0.00	0.45 ± 0.06
Meta.data	0.08 ± 0.04	0.03 ± 0.01
Morphological	0.71 ± 0.00	0.70 ± 0.02
mb-promoters	0.88 ± 0.02	0.76 ± 0.03
Postoperative-patient	0.70 ± 0.02	0.67 ± 0.05
Shuttle-control	0.60 ± 0.02	0.51 ± 0.10
Soybean	0.79 ± 0.06	0.66 ± 0.30
Tempdiag	1.00 ± 0.00	0.97 ± 0.05
Tep.fea	0.65 ± 0.00	0.65 ± 0.00
Tic-tac-toe	0.76 ± 0.04	0.83 ± 0.05
Transfusion	0.79 ± 0.01	0.77 ± 0.02
Vehicle	0.74 ± 0.00	0.70 ± 0.03
Vote	0.95 ± 0.01	0.95 ± 0.01
Vowel	0.59 ± 0.08	0.63 ± 0.09
Wine-red	0.59 ± 0.01	0.57 ± 0.01
Wine-white	0.52 ± 0.01	0.53 ± 0.02
Number of victories	30	4
Wilcoxon p-value	3.04×10^{-6}	

(continued)

Table 5.25   (continued)

(b) *F-Measure results*

Data set	HEAD-DT	Random search
Abalone	0.23 ± 0.00	0.77 ± 0.01
Anneal	0.97 ± 0.01	0.09 ± 0.14
Arrhythmia	0.58 ± 0.17	0.53 ± 0.10
Audiology	0.70 ± 0.02	0.50 ± 0.22
Autos	0.73 ± 0.08	0.39 ± 0.19
Breast-cancer	0.72 ± 0.00	0.36 ± 0.05
Breast-w	0.94 ± 0.01	0.06 ± 0.01
Bridges1	0.68 ± 0.06	0.50 ± 0.13
Bridges2	0.68 ± 0.06	0.51 ± 0.12
Car	0.83 ± 0.02	0.12 ± 0.04
Heart-c	0.81 ± 0.00	0.24 ± 0.04
Flags	0.68 ± 0.01	0.48 ± 0.21
Credit-g	0.73 ± 0.00	0.32 ± 0.06
Colic	0.72 ± 0.18	0.37 ± 0.15
Haberman	0.75 ± 0.01	0.31 ± 0.02
Heart-h	0.80 ± 0.01	0.27 ± 0.11
Ionosphere	0.92 ± 0.03	0.12 ± 0.02
Iris	0.96 ± 0.00	0.05 ± 0.01
kdd-synthetic	0.95 ± 0.01	0.12 ± 0.03
kr-vs-kp	0.91 ± 0.03	0.16 ± 0.27
Liver-disorders	0.72 ± 0.02	0.33 ± 0.02
Lung-cancer	0.69 ± 0.00	0.64 ± 0.10
Meta.data	0.06 ± 0.03	0.98 ± 0.01
Morphological	0.70 ± 0.00	0.31 ± 0.01
mb-promoters	0.88 ± 0.02	0.24 ± 0.03
Postoperative-patient	0.67 ± 0.02	0.42 ± 0.02
Shuttle-control	0.58 ± 0.02	0.60 ± 0.05
Soybean	0.76 ± 0.07	0.37 ± 0.33
Tempdiag	1.00 ± 0.00	0.03 ± 0.05
Tep.fea	0.61 ± 0.00	0.39 ± 0.00
Tic-tac-toe	0.76 ± 0.04	0.18 ± 0.05
Transfusion	0.77 ± 0.00	0.25 ± 0.01
Vehicle	0.74 ± 0.00	0.30 ± 0.03
Vote	0.95 ± 0.01	0.05 ± 0.01
Vowel	0.58 ± 0.09	0.37 ± 0.09
Wine-red	0.57 ± 0.01	0.44 ± 0.02
Wine-white	0.48 ± 0.02	0.49 ± 0.03
Number of victories	33	4
Wilcoxon p-value	8.27×10^{-4}	

Table 5.26 HEAD-DT versus random search: $\{5 \times 35\}$ configuration

(a) *Accuracy results*

Data set	HEAD-DT	Random search
Abalone	0.27 ± 0.00	0.25 ± 0.01
Anneal	0.97 ± 0.00	0.93 ± 0.10
Arrhythmia	0.58 ± 0.08	0.58 ± 0.04
Audiology	0.76 ± 0.00	0.56 ± 0.19
Autos	0.67 ± 0.06	0.64 ± 0.16
Breast-cancer	0.75 ± 0.00	0.68 ± 0.04
Breast-w	0.93 ± 0.00	0.94 ± 0.01
Bridges1	0.64 ± 0.03	0.56 ± 0.11
Bridges2	0.64 ± 0.03	0.56 ± 0.11
Car	0.82 ± 0.02	0.88 ± 0.04
Heart-c	0.81 ± 0.01	0.76 ± 0.03
Flags	0.68 ± 0.01	0.56 ± 0.15
Credit-g	0.75 ± 0.00	0.72 ± 0.01
Colic	0.68 ± 0.09	0.70 ± 0.10
Heart-h	0.77 ± 0.05	0.76 ± 0.07
Ionosphere	0.89 ± 0.00	0.89 ± 0.02
kdd-synthetic	0.96 ± 0.00	0.88 ± 0.03
kr-vs-kp	0.95 ± 0.00	0.87 ± 0.20
Liver-disorders	0.74 ± 0.01	0.67 ± 0.02
Lung-cancer	0.69 ± 0.00	0.45 ± 0.06
Meta.data	0.04 ± 0.02	0.03 ± 0.01
Morphological	0.70 ± 0.00	0.70 ± 0.02
mb-promoters	0.86 ± 0.01	0.76 ± 0.03
Postoperative-patient	0.72 ± 0.02	0.67 ± 0.05
Shuttle-control	0.61 ± 0.01	0.51 ± 0.10
Soybean	0.72 ± 0.02	0.66 ± 0.30
Tempdiag	1.00 ± 0.00	0.97 ± 0.05
Tep.fea	0.65 ± 0.00	0.65 ± 0.00
Tic-tac-toe	0.73 ± 0.03	0.83 ± 0.05
Transfusion	0.79 ± 0.00	0.77 ± 0.02
Vehicle	0.74 ± 0.00	0.70 ± 0.03
Vote	0.96 ± 0.00	0.95 ± 0.01
Vowel	0.50 ± 0.01	0.63 ± 0.09
Wine-red	0.60 ± 0.00	0.57 ± 0.01
Wine-white	0.54 ± 0.00	0.53 ± 0.02
Number of victories	28	6
Wilcoxon p-value	2.25×10^{-4}	

(continued)

Table 5.26 (continued)

(b) *F-Measure results*

Data set	HEAD-DT	Random search
Abalone	0.24 ± 0.00	0.77 ± 0.01
Anneal	0.97 ± 0.00	0.09 ± 0.14
Arrhythmia	0.45 ± 0.13	0.53 ± 0.10
Audiology	0.73 ± 0.00	0.50 ± 0.22
Autos	0.67 ± 0.06	0.39 ± 0.19
Breast-cancer	0.73 ± 0.00	0.36 ± 0.05
Breast-w	0.93 ± 0.00	0.06 ± 0.01
Bridges1	0.63 ± 0.03	0.50 ± 0.13
Bridges2	0.63 ± 0.04	0.51 ± 0.12
Car	0.81 ± 0.02	0.12 ± 0.04
Heart-c	0.80 ± 0.01	0.24 ± 0.04
Flags	0.67 ± 0.01	0.48 ± 0.21
Credit-g	0.73 ± 0.00	0.32 ± 0.06
Colic	0.57 ± 0.14	0.37 ± 0.15
Heart-h	0.74 ± 0.07	0.27 ± 0.11
Ionosphere	0.89 ± 0.01	0.12 ± 0.02
kdd-synthetic	0.96 ± 0.00	0.12 ± 0.03
kr-vs-kp	0.95 ± 0.00	0.16 ± 0.27
Liver-disorders	0.73 ± 0.01	0.33 ± 0.02
Lung-cancer	0.69 ± 0.00	0.64 ± 0.10
Meta.data	0.02 ± 0.01	0.98 ± 0.01
Morphological	0.69 ± 0.01	0.31 ± 0.01
mb-promoters	0.86 ± 0.01	0.24 ± 0.03
Postoperative-patient	0.69 ± 0.03	0.42 ± 0.02
Shuttle-control	0.57 ± 0.02	0.60 ± 0.05
Soybean	0.68 ± 0.01	0.37 ± 0.33
Tempdiag	1.00 ± 0.00	0.03 ± 0.05
Tep.fea	0.61 ± 0.00	0.39 ± 0.00
Tic-tac-toe	0.72 ± 0.04	0.18 ± 0.05
Transfusion	0.77 ± 0.00	0.25 ± 0.01
Vehicle	0.74 ± 0.00	0.30 ± 0.03
Vote	0.96 ± 0.00	0.05 ± 0.01
Vowel	0.48 ± 0.02	0.37 ± 0.09
Wine-red	0.59 ± 0.00	0.44 ± 0.02
Wine-white	0.51 ± 0.00	0.49 ± 0.03
Number of victories	31	4
Wilcoxon p-value	7.84×10^{-6}	

Table 5.27 HEAD-DT versus random search: $\{7 \times 33\}$ configuration

(a) *Accuracy results*

Data set	HEAD-DT	Random search
Abalone	0.29 ± 0.01	0.25 ± 0.01
Anneal	0.98 ± 0.01	0.93 ± 0.10
Arrhythmia	0.78 ± 0.02	0.58 ± 0.04
Audiology	0.79 ± 0.00	0.56 ± 0.19
Autos	0.84 ± 0.04	0.64 ± 0.16
Breast-cancer	0.75 ± 0.00	0.68 ± 0.04
Breast-w	0.96 ± 0.01	0.94 ± 0.01
Bridges2	0.71 ± 0.01	0.56 ± 0.11
Car	0.92 ± 0.02	0.88 ± 0.04
Heart-c	0.83 ± 0.01	0.76 ± 0.03
Flags	0.73 ± 0.01	0.56 ± 0.15
Credit-g	0.76 ± 0.00	0.72 ± 0.01
Colic	0.88 ± 0.01	0.70 ± 0.10
Heart-h	0.83 ± 0.01	0.76 ± 0.07
Ionosphere	0.94 ± 0.01	0.89 ± 0.02
kdd-synthetic	0.95 ± 0.00	0.88 ± 0.03
kr-vs-kp	0.96 ± 0.00	0.87 ± 0.20
Liver-disorders	0.75 ± 0.00	0.67 ± 0.02
Lung-cancer	0.71 ± 0.02	0.45 ± 0.06
Meta.data	0.14 ± 0.02	0.03 ± 0.01
Morphological	0.74 ± 0.01	0.70 ± 0.02
mb-promoters	0.89 ± 0.01	0.76 ± 0.03
Shuttle-control	0.59 ± 0.02	0.51 ± 0.10
Soybean	0.82 ± 0.08	0.66 ± 0.30
Tempdiag	1.00 ± 0.00	0.97 ± 0.05
Tep.fea	0.65 ± 0.00	0.65 ± 0.00
Tic-tac-toe	0.94 ± 0.02	0.83 ± 0.05
Transfusion	0.79 ± 0.00	0.77 ± 0.02
Vehicle	0.77 ± 0.01	0.70 ± 0.03
Vote	0.96 ± 0.00	0.95 ± 0.01
Vowel	0.76 ± 0.07	0.63 ± 0.09
Wine-red	0.64 ± 0.01	0.57 ± 0.01
Wine-white	0.55 ± 0.01	0.53 ± 0.02
Number of victories	32	0
Wilcoxon p-value	5.64×10^{-7}	

(continued)

Table 5.27 (continued)

(b) *F-Measure results*

Data set	HEAD-DT	Random search
Abalone	0.27 ± 0.01	0.77 ± 0.01
Anneal	0.98 ± 0.01	0.09 ± 0.14
Arrhythmia	0.76 ± 0.02	0.53 ± 0.10
Audiology	0.77 ± 0.01	0.50 ± 0.22
Autos	0.85 ± 0.04	0.39 ± 0.19
Breast-cancer	0.73 ± 0.00	0.36 ± 0.05
Breast-w	0.96 ± 0.01	0.06 ± 0.01
Bridges2	0.70 ± 0.02	0.51 ± 0.12
Car	0.92 ± 0.03	0.12 ± 0.04
Heart-c	0.83 ± 0.01	0.24 ± 0.04
Flags	0.72 ± 0.01	0.48 ± 0.21
Credit-g	0.76 ± 0.01	0.32 ± 0.06
Colic	0.88 ± 0.01	0.37 ± 0.15
Heart-h	0.83 ± 0.01	0.27 ± 0.11
Ionosphere	0.94 ± 0.01	0.12 ± 0.02
kdd-synthetic	0.95 ± 0.00	0.12 ± 0.03
kr-vs-kp	0.96 ± 0.00	0.16 ± 0.27
Liver-disorders	0.75 ± 0.00	0.33 ± 0.02
Lung-cancer	0.71 ± 0.02	0.64 ± 0.10
Meta.data	0.13 ± 0.02	0.98 ± 0.01
Morphological	0.72 ± 0.00	0.31 ± 0.01
mb-promoters	0.89 ± 0.01	0.24 ± 0.03
Shuttle-control	0.55 ± 0.02	0.60 ± 0.05
Soybean	0.80 ± 0.09	0.37 ± 0.33
Tempdiag	1.00 ± 0.00	0.03 ± 0.05
Tep.fea	0.61 ± 0.00	0.39 ± 0.00
Tic-tac-toe	0.94 ± 0.02	0.18 ± 0.05
Transfusion	0.77 ± 0.00	0.25 ± 0.01
Vehicle	0.77 ± 0.01	0.30 ± 0.03
Vote	0.96 ± 0.00	0.05 ± 0.01
Vowel	0.76 ± 0.07	0.37 ± 0.09
Wine-red	0.63 ± 0.01	0.44 ± 0.02
Wine-white	0.53 ± 0.01	0.49 ± 0.03
Number of victories	30	3
Wilcoxon p-value	6.36×10^{-6}	

Table 5.28 HEAD-DT vs random search: $\{9 \times 31\}$ configuration

(a) *Accuracy results*

Data set	HEAD-DT	Random search
Abalone	0.28 ± 0.00	0.25 ± 0.01
Anneal	0.92 ± 0.00	0.93 ± 0.10
Arrhythmia	0.76 ± 0.00	0.58 ± 0.04
Audiology	0.67 ± 0.01	0.56 ± 0.19
Autos	0.78 ± 0.01	0.64 ± 0.16
Breast-cancer	0.73 ± 0.00	0.68 ± 0.04
Breast-w	0.94 ± 0.00	0.94 ± 0.01
Bridges2	0.62 ± 0.01	0.56 ± 0.11
Car	0.88 ± 0.01	0.88 ± 0.04
Heart-c	0.82 ± 0.00	0.76 ± 0.03
Flags	0.71 ± 0.01	0.56 ± 0.15
Credit-g	0.75 ± 0.00	0.72 ± 0.01
Colic	0.83 ± 0.01	0.70 ± 0.10
Heart-h	0.80 ± 0.00	0.76 ± 0.07
Ionosphere	0.92 ± 0.00	0.89 ± 0.02
kdd-synthetic	0.95 ± 0.00	0.88 ± 0.03
kr-vs-kp	0.91 ± 0.01	0.87 ± 0.20
Liver-disorders	0.74 ± 0.00	0.67 ± 0.02
Meta.data	0.11 ± 0.00	0.03 ± 0.01
Morphological	0.71 ± 0.00	0.70 ± 0.02
mb-promoters	0.73 ± 0.00	0.76 ± 0.03
Shuttle-control	0.63 ± 0.00	0.51 ± 0.10
Soybean	0.64 ± 0.01	0.66 ± 0.30
Tep.fea	0.65 ± 0.00	0.65 ± 0.00
Tic-tac-toe	0.83 ± 0.01	0.83 ± 0.05
Transfusion	0.79 ± 0.00	0.77 ± 0.02
Vehicle	0.75 ± 0.00	0.70 ± 0.03
Vote	0.96 ± 0.00	0.95 ± 0.01
Vowel	0.61 ± 0.01	0.63 ± 0.09
Wine-red	0.61 ± 0.00	0.57 ± 0.01
Wine-white	0.53 ± 0.00	0.53 ± 0.02
Number of victories	26	4
Wilcoxon p-value	2.51×10^{-5}	

(b) *F-Measure results*

Data set	HEAD-DT	Random search
Abalone	0.24 ± 0.00	0.77 ± 0.01
Anneal	0.92 ± 0.00	0.09 ± 0.14
Arrhythmia	0.73 ± 0.00	0.53 ± 0.10

(continued)

Table 5.28 (continued)

(b) *F-Measure results*

Data set	HEAD-DT	Random search
Audiology	0.62 ± 0.01	0.50 ± 0.22
Autos	0.78 ± 0.01	0.39 ± 0.19
Breast-cancer	0.71 ± 0.00	0.36 ± 0.05
Breast-w	0.94 ± 0.00	0.06 ± 0.01
Bridges2	0.60 ± 0.01	0.51 ± 0.12
Car	0.88 ± 0.01	0.12 ± 0.04
Heart-c	0.82 ± 0.00	0.24 ± 0.04
Flags	0.70 ± 0.00	0.48 ± 0.21
Credit-g	0.73 ± 0.00	0.32 ± 0.06
Colic	0.83 ± 0.01	0.37 ± 0.15
Heart-h	0.79 ± 0.00	0.27 ± 0.11
Ionosphere	0.92 ± 0.00	0.12 ± 0.02
kdd-synthetic	0.95 ± 0.00	0.12 ± 0.03
kr-vs-kp	0.91 ± 0.01	0.16 ± 0.27
Liver-disorders	0.73 ± 0.00	0.33 ± 0.02
Meta.data	0.09 ± 0.00	0.98 ± 0.01
Morphological	0.69 ± 0.00	0.31 ± 0.01
mb-promoters	0.73 ± 0.00	0.24 ± 0.03
Shuttle-control	0.56 ± 0.00	0.60 ± 0.05
Soybean	0.60 ± 0.02	0.37 ± 0.33
Tep.fea	0.61 ± 0.00	0.39 ± 0.00
Tic-tac-toe	0.83 ± 0.02	0.18 ± 0.05
Transfusion	0.77 ± 0.00	0.25 ± 0.01
Vehicle	0.74 ± 0.00	0.30 ± 0.03
Vote	0.96 ± 0.00	0.05 ± 0.01
Vowel	0.60 ± 0.01	0.37 ± 0.09
Wine-red	0.59 ± 0.00	0.44 ± 0.02
Wine-white	0.49 ± 0.00	0.49 ± 0.03
Number of victories	28	3
Wilcoxon *p*-value	4.01×10^{-5}	

Even though HEAD-DT has already presented quite satisfactory results, we investigate in Chap. 6 several different strategies for using as HEAD-DT's fitness function in the general framework. We test both balanced and imbalanced data in the meta-training set used in the evolutionary process.

References

1. R.C. Barros et al., Automatic design of decision-tree induction algorithms tailored to flexible-receptor docking data, in *BMC Bioinformatics* 13 (2012)
2. R.C. Barros et al., Towards the automatic design of decision tree induction algorithms, in *13th Annual Conference Companion on Genetic and Evolutionary Computation* (GECCO 2011). pp. 567–574 (2011)
3. M.P. Basgalupp et al., Software effort prediction: a hyper-heuristic decision-tree based approach, in *28th Annual ACM Symposium on Applied Computing*. pp. 1109–1116 (2013)
4. L. Breiman et al., *Classification and Regression Trees* (Wadsworth, Belmont, 1984)
5. B. Chandra, R. Kothari, P. Paul, A new node splitting measure for decision tree construction. Pattern Recognit. **43**(8), 2725–2731 (2010)
6. B. Chandra, P.P. Varghese, Moving towards efficient decision tree construction. Inf. Sci. **179**(8), 1059–1069 (2009)
7. J. Demšar, Statistical comparisons of classifiers over multiple data sets. J. Mach. Learn. Res. **7**, 1–30 (2006). ISSN: 1532–4435
8. A. Frank, A. Asuncion, *UCI Machine Learning Repository* (2010)
9. R. Iman, J. Davenport, Approximations of the critical region of the Friedman statistic, in *Communications in Statistics*, pp. 571–595 (1980)
10. S. Monti et al., Consensus clustering: a resampling-based method for class discovery and visualization of gene expression microarray data. Mach. Learn. **52**(1–2), 91–118 (2003)
11. J.R. Quinlan, *C4.5: Programs for Machine Learning* (Morgan Kaufmann, San Francisco, 1993). ISBN: 1-55860-238-0
12. M. Souto et al., Clustering cancer gene expression data: a comparative study. BMC Bioinform. **9**(1), 497 (2008)
13. F. Wilcoxon, Individual comparisons by ranking methods. Biometrics **1**, 80–83 (1945)
14. I.H. Witten, E. Frank, *Data Mining: Practical Machine Learning Tools and Techniques with Java Implementations* (Morgan Kaufmann, San Francisco, 1999). ISBN: 1558605525

Chapter 6
HEAD-DT: Fitness Function Analysis

Abstract In Chap. 4, more specifically in Sect. 4.4, we saw that the definition of a fitness function for the scenario in which HEAD-DT evolves a decision-tree algorithm from multiple data sets is an interesting and relevant problem. In the experiments presented in Chap. 5, Sect. 5.2, we employed a simple average over the F-Measure obtained in the data sets that belong to the meta-training set. As previously observed, when evolving an algorithm from multiple data sets, each individual of HEAD-DT has to be executed over each data set in the meta-training set. Hence, instead of obtaining a single value of predictive performance, each individual scores a set of values that have to be eventually combined into a single measure. In this chapter, we analyse in more detail the impact of different strategies to be used as fitness function during the evolutionary cycle of HEAD-DT. We divide the experimental scheme into two distinct scenarios: (i) evolving a decision-tree induction algorithm from multiple balanced data sets; and (ii) evolving a decision-tree induction algorithm from multiple imbalanced data sets. In each of these scenarios, we analyse the difference in performance of well-known performance measures such as accuracy, F-Measure, AUC, recall, and also a lesser-known criterion, namely the relative accuracy improvement. In addition, we analyse different schemes of aggregation, such as simple average, median, and harmonic mean.

Keywords Fitness functions · Performance measures · Evaluation schemes

6.1 Performance Measures

Performance measures are key tools to assess the quality of machine learning approaches and models. Therefore, several different measures have been proposed in the specialized literature with the goal of providing better choices in general or for a specific application domain [2].

In the context of HEAD-DT's fitness function, and given that it evaluates algorithms (individuals) over data sets, it is reasonable to assume that different

---

Assuming we are not interested in dealing with a multi-objective optimisation problem.

© The Author(s) 2015
R.C. Barros et al., *Automatic Design of Decision-Tree Induction Algorithms*,
SpringerBriefs in Computer Science, DOI 10.1007/978-3-319-14231-9_6

classification performance measures could be employed to provide a quantitative assessment of algorithmic performance. In the next few sections, we present five different performance measures that were selected for further investigation as HEAD-DT's fitness function.

6.1.1 Accuracy

Probably the most well-known performance evaluation measure for classification problems, the accuracy of a model is the rate of correctly classified instances:

$$accuracy = \frac{tp + tn}{tp + tn + fp + fn} \tag{6.1}$$

where $tp(tn)$ stands for the true positives (true negatives)—instances correctly classified,—and $fp(fn)$ stands for the false positives (false negatives)—instances incorrectly classified.

Even though most classification algorithms are assessed regarding the accuracy they obtain in a data set, we must point out that accuracy may be a misleading performance measure. For instance, suppose we have a data set whose class distribution is very skewed: 90 % of the instances belong to class A and 10 % to class B. An algorithm that always classifies instances as belonging to class A would achieve 90 % of accuracy, even though it never predicts a class-B instance. In this case, assuming that class B is equally important (or even more so) than class A, we would prefer an algorithm with lower accuracy, but which could eventually correctly predict some instances as belonging to the rare class B.

6.1.2 F-Measure

As it was presented in Sect. 4.4, F-Measure (also F-score or F_1 score) is the harmonic mean of precision and recall:

$$precision = \frac{tp}{tp + fp} \tag{6.2}$$

$$recall = \frac{tp}{tp + fn} \tag{6.3}$$

$$f1 = 2 \times \frac{precision \times recall}{precision + recall} \tag{6.4}$$

Note that though F-Measure is advocated in the machine learning literature as a single measure capable of capturing the effectiveness of a system, it still completely ignores the tn, which can vary freely without affecting the statistic [8].

6.1.3 Area Under the ROC Curve

The area under the ROC (receiver operating characteristic) curve (AUC) has been increasingly used as a performance evaluation measure in classification problems. The ROC curve graphically displays the trade-off between the true positive rate ($tpr = tp/(tp + fn)$) and the false positive rate ($fpr = fp/(fp + tn)$) of a classifier. ROC graphs have properties that make them especially useful for domains with skewed class distribution and unequal classification error costs [1].

To create the ROC curve, one needs to build a graph in which the tpr is plotted along the y axis and the fpr is shown on the x axis. Each point along the curve corresponds to one of the models induced by a given algorithm, and different models are built by varying a probabilistic threshold that determines whether an instance should be classified as positive or negative.

A ROC curve is a two-dimensional depiction of a classifier. To compare classifiers, we may want to reduce ROC performance to a single scalar value representing the expected performance, which is precisely the AUC. Since the AUC is a portion of the area of the unit square, its value will always be between 0 and 1. However, because random guessing produces a diagonal line between (0,0) and (1,1), which has an area of 0.5, no realistic classifier should have an AUC value of less than 0.5. The AUC has an important statistical property: it is equivalent to the probability that the classifier will rank a randomly chosen positive instance higher than a randomly chosen negative instance, which makes of the AUC equivalent to the Wilcoxon test of ranks [6].

The machine learning community often uses the AUC statistic for model comparison, even though this practice has recently been questioned based upon new research that shows that AUC is quite noisy as a performance measure for classification [3] and has some other significant problems in model comparison [4, 5].

6.1.4 Relative Accuracy Improvement

Originally proposed by Pappa [7], the relative accuracy improvement criterion measures the normalized improvement in accuracy of a given model over the data set's default accuracy (i.e., the accuracy achieved when using the majority class of the training data to classify the unseen data):

$$RAI_i = \begin{cases} \frac{Acc_i - DefAcc_i}{1-DefAcc_i}, & \text{if } Acc_i > DefAcc_i \\ \frac{Acc_i - DefAcc_i}{DefAcc_i}, & \text{otherwise} \end{cases} \quad (6.5)$$

In Eq. (6.5), Acc_i is the accuracy achieved by a given classifier in data set i, whereas $DefAcc_i$ is the default accuracy of data set i. Note that if the improvement in accuracy is positive, i.e., the classifier accuracy is greater than the default accuracy, the improvement is normalized by the maximum possible improvement over the default accuracy $(1 - DefAcc_i)$. Otherwise, the drop in the accuracy is normalized by the maximum possible drop, which is the value of the default accuracy itself. Hence, the relative accuracy improvement RAI_i regarding data set i returns a value between -1 (when $Acc_i = 0$) and 1 (when $Acc_i = 1$). Any improvement regarding the default accuracy results in a positive value, whereas any drop results in a negative value. In case $Acc_i = DefAcc_i$ (i.e., no improvement or drop in accuracy is achieved), $RAI_i = 0$, as expected.

The disadvantage of the relative accuracy improvement criterion is that it is not suitable for very imbalanced problems—data sets in which the default accuracy is really close to 1,—since high accuracy does not properly translate into high performance for these kinds of problems, as we have previously seen.

6.1.5 Recall

Also known as *sensitivity* (usually in the medical field) or *true positive rate*, recall measures the proportion of actual positives that are correctly identified as such. For instance, it may refer to the percentage of sick patients who are correctly classified as having the particular disease. In terms of the confusion matrix terms, recall is computed as follows:

$$recall = \frac{tp}{tp + fn} \quad (6.6)$$

Recall is useful for the case of imbalanced data, in which the positive class is the rare class. However, note that a classifier that always predicts the positive class will achieve a perfect recall, since recall does not take into consideration the fp values. This problem is alleviated in multi-class problems, in which each class is used in turn as the positive class, and the average of the per-class recall is taken.

6.2 Aggregation Schemes

All classification measures presented in the previous section refer to the predictive performance of a given classifier in a given data set. When evolving an algorithm from multiple data sets, HEAD-DT's fitness function is measured as the aggregated performance of the individual in each data set that belongs to the meta-training set. We propose employing three simple strategies for combining the per-data-set performance into a single quantitative value: (i) simple average; (ii) median; and (iii) harmonic mean.

The simple average (or alternatively the arithmetic average) is computed by simply taking the average of the per-data-set values, i.e., $(1/N) \times \sum_{i=1}^{N} p_i$, for a meta-training set with N data sets and a performance measure p. It gives equal importance to the performance achieved in each data set. Moreover, it is best used in situations where there are no extreme outliers and the values are independent of each other.

The median is computed by ordering the performance values from smallest to greatest, and then taking the middle value of the ordered list. If there is an even number of data sets, since there is no single middle value, either $N/2$ or $(N/2)+1$ can be used as middle value, or alternatively their average. The median is robust to outliers in the data (extremely large or extremely low values that may influence the simple average).

Finally, the harmonic mean is given by $\left((1/N) \times \sum_{i=1}^{N} p_i\right)^{-1}$. Unlike the simple average, the harmonic mean gives less significance to high-value outliers, providing sometimes a better picture of the average.

6.3 Experimental Evaluation

In this section, we perform an empirical evaluation of the five classification performance measures presented in Sect. 6.1 and the three aggregation schemes presented in Sect. 6.2 as fitness functions of HEAD-DT. There are a total of 15 distinct fitness functions resulting from this analysis:

1. Accuracy + Simple Average (ACC-A);
2. Accuracy + Median (ACC-M);
3. Accuracy + Harmonic Mean (ACC-H);
4. AUC + Simple Average (AUC-A);
5. AUC + Median (AUC-M);
6. AUC + Harmonic Mean (AUC-H);
7. F-Measure + Simple Average (FM-A);
8. F-Measure + Median (FM-M);
9. F-Measure + Harmonic Mean (FM-H);
10. Relative Accuracy Improvement + Simple Average (RAI-A);
11. Relative Accuracy Improvement + Median (RAI-M);
12. Relative Accuracy Improvement + Harmonic Mean (RAI-H);
13. Recall + Simple Average (TPR-A);
14. Recall + Median (TPR-M);
15. Recall + Harmonic Mean (TPR-H).

For this experiment, we employed the 67 UCI data sets described in Table 5.14 organized into two scenarios: (i) 5 balanced data sets in the meta-training set; and (ii) 5 imbalanced data sets in the training set. These scenarios were created to assess the performance of the 15 distinct fitness functions in balanced and imbalanced data, considering that some of the performance measures are explicitly designed to deal with imbalanced data whereas others are not. The term "(im)balanced" was quantitatively measured according to the imbalance ratio (IR):

$$IR = \frac{F(A_{DS})}{F(B_{DS})} \tag{6.7}$$

where $F(.)$ returns the frequency of a given class, A_{DS} is the highest-frequency class in data set DS and B_{DS} the lowest-frequency class in data set DS.

Given the size and complexity of this experiment, we did not optimise HEAD-DT's parameters as in Chap. 5, Sect. 5.2. Instead, we employed typical values found in the literature of evolutionary algorithms for decision-tree induction (the same parameters as in Chap. 5, Sect. 5.1):

- Population size: 100;
- Maximum number of generations: 100;
- Selection: tournament selection with size $t = 2$;
- Elitism rate: 5 individuals;
- Crossover: uniform crossover with 90 % probability;
- Mutation: random uniform gene mutation with 5 % probability.
- Reproduction: cloning individuals with 5 % probability.

In the next sections, we present the results for both scenarios of meta-training set. Moreover, in the end of this chapter, we perform a whole new set of experiments with the best-performing fitness functions.

6.3.1 Results for the Balanced Meta-Training Set

We randomly selected 5 balanced data sets ($IR < 1.1$) from the 67 UCI data sets described in Table 5.14 to be part of the meta-training set in this experiment: iris ($IR = 1.0$), segment ($IR = 1.0$), vowel ($IR = 1.0$), mushroom ($IR = 1.07$), and kr-vs-kp ($IR = 1.09$).

Tables 6.1 and 6.2 show the results for the 62 data sets in the meta-test set regarding accuracy and F-Measure, respectively. At the bottom of each table, the average rank is presented for the 15 versions of HEAD-DT created by varying the fitness functions. We did not present standard deviation values due to space limitations within the tables.

By careful inspection of both tables, we can see that their rankings are practically the same, with the median of the relative accuracy improvement being the

Table 6.1 Accuracy values for the 15 versions of HEAD-DT varying the fitness functions

	ACC A	ACC M	ACC H	AUC A	AUC M	AUC H	FM A	FM M	FM H	RAI A	RAI M	RAI H	TPR A	TPR M	TPR H
Abalone	0.48	0.55	0.50	0.38	0.44	0.34	0.50	0.56	0.50	0.54	0.57	0.53	0.59	0.56	0.50
Anneal	0.99	0.99	0.99	0.99	0.99	0.99	0.99	1.00	0.99	1.00	1.00	0.99	1.00	1.00	1.00
Arrhythmia	0.85	0.85	0.82	0.78	0.78	0.75	0.82	0.85	0.80	0.85	0.86	0.80	0.84	0.73	0.82
Audiology	0.86	0.86	0.86	0.79	0.80	0.79	0.85	0.86	0.86	0.85	0.88	0.82	0.87	0.86	0.84
Autos	0.85	0.83	0.85	0.84	0.71	0.83	0.85	0.82	0.85	0.86	0.88	0.87	0.87	0.78	0.85
Balance-scale	0.78	0.76	0.77	0.77	0.75	0.76	0.78	0.81	0.77	0.81	0.81	0.85	0.78	0.79	0.77
Breast-cancer	0.63	0.75	0.63	0.63	0.69	0.68	0.66	0.74	0.66	0.66	0.73	0.78	0.70	0.77	0.67
Breast-w	0.99	0.99	0.99	0.99	0.99	0.98	0.99	0.99	0.99	0.98	0.99	0.97	0.99	0.99	0.98
Bridges1	0.63	0.73	0.59	0.61	0.73	0.65	0.62	0.74	0.61	0.66	0.74	0.60	0.64	0.74	0.64
Bridges2	0.95	0.98	0.95	0.94	0.98	0.93	0.95	0.98	0.95	0.95	0.98	0.61	0.96	0.99	0.95
Car	0.85	0.82	0.85	0.83	0.83	0.82	0.85	0.86	0.85	0.85	0.87	0.95	0.86	0.85	0.85
cmc	0.88	0.90	0.88	0.88	0.89	0.88	0.88	0.89	0.88	0.89	0.89	0.67	0.89	0.90	0.88
Colic	0.84	0.85	0.84	0.84	0.84	0.83	0.81	0.83	0.80	0.85	0.85	0.87	0.79	0.82	0.76
Column-2C	0.89	0.87	0.89	0.87	0.87	0.87	0.89	0.89	0.89	0.89	0.89	0.88	0.90	0.87	0.89
Column-3C	0.89	0.88	0.89	0.88	0.89	0.88	0.89	0.90	0.89	0.89	0.90	0.89	0.89	0.89	0.89
Credit-a	0.77	0.82	0.82	0.77	0.79	0.73	0.79	0.83	0.80	0.77	0.84	0.89	0.81	0.80	0.71
Credit-g	0.83	0.81	0.83	0.80	0.81	0.79	0.83	0.84	0.83	0.83	0.84	0.81	0.84	0.84	0.84
Cylinder-bands	0.96	0.96	0.96	0.95	0.95	0.95	0.96	0.96	0.96	0.96	0.96	0.83	0.96	0.96	0.96
Dermatology	0.89	0.88	0.89	0.86	0.87	0.86	0.89	0.89	0.89	0.89	0.89	0.96	0.89	0.87	0.89
Diabetes	0.51	0.56	0.50	0.47	0.50	0.48	0.47	0.55	0.49	0.53	0.52	0.84	0.54	0.55	0.47
Ecoli	0.78	0.79	0.76	0.75	0.76	0.74	0.76	0.79	0.76	0.78	0.79	0.89	0.78	0.81	0.76

(continued)

Table 6.1 (continued)

	ACC			AUC			FM			RAI			TPR		
	A	M	H	A	M	H	A	M	H	A	M	H	A	M	H
Flags	0.81	0.81	0.80	0.79	0.81	0.78	0.81	0.84	0.80	0.81	0.84	0.78	0.82	0.82	0.80
Glass	0.79	0.77	0.79	0.78	0.78	0.77	0.80	0.80	0.80	0.80	0.81	0.83	0.81	0.79	0.80
Haberman	0.86	0.86	0.86	0.87	0.87	0.86	0.86	0.87	0.86	0.87	0.87	0.79	0.86	0.87	0.86
Hayes-roth	0.85	0.84	0.85	0.84	0.84	0.84	0.85	0.87	0.85	0.86	0.87	0.86	0.86	0.87	0.85
Heart-c	0.66	0.69	0.66	0.61	0.63	0.60	0.66	0.70	0.66	0.68	0.70	0.85	0.69	0.70	0.66
Heart-h	0.94	0.94	0.94	0.93	0.93	0.93	0.94	0.94	0.94	0.94	0.94	0.84	0.95	0.94	0.94
Heart-statlog	0.87	0.84	0.89	0.87	0.85	0.86	0.88	0.87	0.86	0.85	0.89	0.85	0.85	0.85	0.84
Hepatitis	0.83	0.85	0.88	0.82	0.83	0.72	0.79	0.83	0.82	0.77	0.88	0.90	0.80	0.80	0.73
Ionosphere	0.81	0.83	0.81	0.86	0.80	0.80	0.80	0.83	0.79	0.77	0.80	0.94	0.76	0.80	0.73
kdd-synthetic	0.79	0.81	0.79	0.77	0.78	0.76	0.79	0.80	0.79	0.80	0.80	0.96	0.82	0.80	0.79
Labor	0.68	0.58	0.68	0.65	0.68	0.66	0.67	0.70	0.67	0.69	0.71	0.87	0.69	0.66	0.67
Liver-disorders	0.85	0.87	0.85	0.83	0.85	0.85	0.85	0.88	0.85	0.86	0.88	0.79	0.85	0.89	0.85
Lung-cancer	0.25	0.23	0.30	0.14	0.17	0.10	0.26	0.32	0.29	0.23	0.35	0.68	0.32	0.28	0.23
Lymph	0.77	0.79	0.78	0.75	0.77	0.75	0.78	0.79	0.78	0.78	0.79	0.85	0.79	0.79	0.78
mb-promoters	0.71	0.72	0.72	0.70	0.70	0.70	0.72	0.73	0.73	0.74	0.73	0.30	0.75	0.72	0.73
Meta.data	0.87	0.86	0.86	0.87	0.87	0.86	0.86	0.89	0.86	0.86	0.89	0.87	0.87	0.88	0.87
Morphological	0.83	0.83	0.83	0.81	0.83	0.80	0.83	0.84	0.83	0.84	0.84	0.78	0.85	0.84	0.83
Postoperative	1.00	1.00	1.00	1.00	1.00	1.00	1.00	1.00	1.00	1.00	1.00	0.71	1.00	1.00	1.00
Primary-tumor	1.00	1.00	1.00	1.00	1.00	1.00	1.00	1.00	1.00	1.00	1.00	0.51	1.00	1.00	1.00
Readings-2	0.77	0.77	0.77	0.74	0.76	0.74	0.77	0.78	0.76	0.77	0.78	1.00	0.78	0.78	0.77
Readings-4	0.77	0.77	0.77	0.76	0.77	0.76	0.77	0.79	0.77	0.78	0.79	1.00	0.78	0.79	0.77

(continued)

Table 6.1 (continued)

	ACC A	ACC M	ACC H	AUC A	AUC M	AUC H	FM A	FM M	FM H	RAI A	RAI M	RAI H	TPR A	TPR M	TPR H
Semeion	0.66	0.60	0.63	0.59	0.55	0.61	0.63	0.66	0.57	0.63	0.63	0.97	0.63	0.63	0.62
Shuttle-control	0.89	0.87	0.90	0.91	0.88	0.91	0.86	0.90	0.89	0.91	0.88	0.67	0.91	0.89	0.90
Sick	0.88	0.84	0.88	0.85	0.84	0.85	0.88	0.87	0.88	0.88	0.87	0.99	0.88	0.83	0.88
Solar-flare-1	0.95	0.94	0.95	0.94	0.93	0.93	0.95	0.96	0.95	0.96	0.96	0.77	0.95	0.95	0.95
Solar-flare-2	0.72	0.75	0.72	0.67	0.64	0.67	0.72	0.76	0.72	0.74	0.77	0.77	0.76	0.75	0.73
Sonar	1.00	1.00	1.00	1.00	1.00	1.00	1.00	1.00	1.00	1.00	1.00	0.88	1.00	1.00	1.00
Soybean	0.65	0.65	0.65	0.65	0.65	0.65	0.65	0.65	0.65	0.65	0.65	0.93	0.65	0.65	0.65
Sponge	0.91	0.95	0.90	0.89	0.95	0.89	0.90	0.94	0.90	0.91	0.94	0.95	0.91	0.96	0.91
Tae	0.87	0.55	0.88	0.78	0.63	0.79	0.77	0.80	0.75	0.88	0.79	0.72	0.67	0.58	0.71
Tempdiag	0.80	0.79	0.80	0.79	0.79	0.79	0.80	0.81	0.80	0.80	0.81	1.00	0.81	0.80	0.80
Tep.fea	0.84	0.86	0.85	0.80	0.82	0.78	0.85	0.86	0.85	0.86	0.86	0.65	0.87	0.85	0.84
Tic-tac-toe	0.97	0.96	0.97	0.97	0.96	0.96	0.96	0.96	0.96	0.97	0.97	0.91	0.96	0.96	0.97
Trains	0.96	0.94	0.96	0.95	0.94	0.95	0.96	0.95	0.96	0.96	0.95	0.85	0.96	0.93	0.96
Transfusion	0.75	0.79	0.77	0.70	0.73	0.67	0.77	0.80	0.77	0.79	0.80	0.80	0.81	0.80	0.76
Vehicle	0.97	0.96	0.97	0.96	0.96	0.95	0.97	0.97	0.96	0.96	0.97	0.86	0.97	0.97	0.96
Vote	0.90	0.91	0.86	0.94	0.95	0.94	0.86	0.96	0.86	0.90	0.91	0.97	0.77	0.96	0.83
Wine	0.62	0.61	0.66	0.62	0.63	0.57	0.64	0.60	0.62	0.57	0.64	0.96	0.62	0.59	0.58
Wine-red	0.93	0.90	0.93	0.90	0.88	0.91	0.93	0.90	0.93	0.93	0.91	0.78	0.93	0.89	0.93
Wine-white	0.94	0.94	0.94	0.93	0.94	0.93	0.94	0.94	0.94	0.94	0.94	0.78	0.94	0.94	0.94
Zoo	0.59	0.61	0.59	0.56	0.58	0.54	0.59	0.62	0.59	0.61	0.61	0.90	0.61	0.61	0.58
Average rank	8.00	8.93	8.35	11.68	10.76	12.57	8.25	4.75	9.10	6.41	**3.72**	6.64	4.93	6.88	9.04

Meta-training comprises 5 balanced data sets

Table 6.2 F-Measure values for the 15 versions of HEAD-DT varying the fitness functions

	ACC A	ACC M	ACC H	AUC A	AUC M	AUC H	FM A	FM M	FM H	RAI A	RAI M	RAI H	TPR A	TPR M	TPR H
Abalone	0.48	0.55	0.50	0.37	0.43	0.33	0.50	0.56	0.50	0.53	0.57	0.53	0.59	0.56	0.49
Anneal	0.99	0.99	0.99	0.99	0.99	0.99	0.99	1.00	0.99	1.00	1.00	0.99	1.00	1.00	1.00
Arrhythmia	0.57	0.56	0.65	0.56	0.58	0.47	0.60	0.53	0.56	0.46	0.62	0.79	0.57	0.52	0.49
Audiology	0.84	0.83	0.80	0.75	0.75	0.73	0.81	0.84	0.78	0.84	0.85	0.81	0.83	0.68	0.81
Autos	0.85	0.86	0.86	0.79	0.79	0.78	0.85	0.86	0.86	0.85	0.88	0.87	0.87	0.86	0.84
Balance-scale	0.85	0.82	0.85	0.82	0.68	0.81	0.85	0.82	0.85	0.87	0.88	0.85	0.87	0.77	0.85
Breast-cancer	0.76	0.71	0.72	0.76	0.74	0.75	0.75	0.80	0.74	0.80	0.81	0.76	0.75	0.76	0.73
Breast-w	0.97	0.96	0.97	0.96	0.96	0.95	0.97	0.97	0.96	0.96	0.97	0.97	0.97	0.97	0.96
Bridges1	0.60	0.74	0.60	0.61	0.69	0.67	0.65	0.73	0.65	0.65	0.73	0.56	0.70	0.77	0.66
Bridges2	0.60	0.72	0.54	0.60	0.72	0.64	0.58	0.73	0.57	0.65	0.73	0.56	0.59	0.72	0.60
Car	0.95	0.98	0.95	0.94	0.98	0.93	0.95	0.98	0.95	0.95	0.98	0.95	0.96	0.99	0.95
cmc	0.66	0.69	0.65	0.61	0.63	0.60	0.66	0.69	0.66	0.68	0.70	0.66	0.69	0.70	0.66
Colic	0.82	0.85	0.88	0.81	0.82	0.67	0.75	0.82	0.81	0.73	0.88	0.87	0.77	0.77	0.68
Column-2C	0.88	0.90	0.88	0.88	0.89	0.88	0.88	0.89	0.88	0.89	0.89	0.88	0.89	0.90	0.88
Column-3C	0.89	0.87	0.89	0.87	0.87	0.87	0.89	0.89	0.89	0.89	0.89	0.89	0.90	0.86	0.89
Credit-a	0.89	0.88	0.89	0.88	0.89	0.88	0.89	0.90	0.89	0.89	0.90	0.89	0.89	0.89	0.89
Credit-g	0.80	0.78	0.80	0.78	0.81	0.78	0.80	0.84	0.80	0.81	0.84	0.80	0.82	0.79	0.80
Cylinder-bands	0.75	0.82	0.82	0.77	0.79	0.72	0.79	0.83	0.80	0.77	0.84	0.83	0.81	0.80	0.69
Dermatology	0.96	0.96	0.96	0.95	0.95	0.95	0.96	0.96	0.96	0.96	0.96	0.96	0.96	0.96	0.96
Diabetes	0.83	0.82	0.83	0.81	0.83	0.80	0.83	0.84	0.83	0.84	0.84	0.84	0.85	0.84	0.83
Ecoli	0.88	0.88	0.88	0.86	0.87	0.86	0.88	0.89	0.88	0.89	0.89	0.88	0.89	0.86	0.88

(continued)

Table 6.2 (continued)

	ACC			AUC			FM			RAI			TPR		
	A	M	H	A	M	H	A	M	H	A	M	H	A	M	H
Flags	0.78	0.79	0.76	0.74	0.75	0.74	0.76	0.79	0.76	0.78	0.79	0.78	0.78	0.81	0.76
Glass	0.83	0.81	0.83	0.79	0.80	0.78	0.83	0.83	0.83	0.83	0.84	0.83	0.84	0.84	0.84
Haberman	0.78	0.73	0.78	0.76	0.77	0.75	0.78	0.79	0.78	0.79	0.80	0.77	0.80	0.76	0.79
Hayes-roth	0.86	0.86	0.86	0.87	0.87	0.86	0.86	0.86	0.86	0.86	0.87	0.86	0.86	0.86	0.86
Heart-c	0.85	0.82	0.85	0.83	0.83	0.82	0.85	0.86	0.85	0.85	0.87	0.85	0.86	0.85	0.85
Heart-h	0.84	0.85	0.84	0.84	0.84	0.83	0.79	0.81	0.78	0.85	0.85	0.83	0.75	0.80	0.72
Heart-statlog	0.85	0.84	0.85	0.84	0.84	0.84	0.85	0.87	0.85	0.86	0.87	0.85	0.86	0.87	0.85
Hepatitis	0.85	0.80	0.89	0.86	0.83	0.83	0.86	0.86	0.84	0.82	0.88	0.89	0.83	0.83	0.82
Ionosphere	0.94	0.94	0.94	0.93	0.93	0.93	0.94	0.94	0.94	0.94	0.94	0.94	0.95	0.94	0.94
kdd-synthetic	0.93	0.90	0.93	0.90	0.88	0.91	0.93	0.90	0.93	0.93	0.91	0.96	0.93	0.88	0.93
Labor	0.78	0.80	0.79	0.85	0.78	0.77	0.78	0.83	0.76	0.72	0.76	0.87	0.71	0.77	0.67
Liver-disorders	0.79	0.81	0.79	0.77	0.78	0.76	0.79	0.80	0.79	0.80	0.80	0.79	0.82	0.80	0.79
Lung-cancer	0.68	0.54	0.68	0.65	0.68	0.66	0.67	0.70	0.67	0.69	0.71	0.68	0.69	0.64	0.67
Lymph	0.85	0.87	0.85	0.83	0.84	0.85	0.85	0.88	0.85	0.86	0.88	0.85	0.85	0.89	0.85
mb-promoters	0.87	0.86	0.86	0.86	0.87	0.86	0.86	0.89	0.86	0.86	0.89	0.86	0.87	0.88	0.86
Meta.data	0.24	0.21	0.28	0.12	0.16	0.08	0.25	0.32	0.28	0.24	0.35	0.28	0.32	0.28	0.23
Morphological	0.77	0.79	0.77	0.75	0.76	0.74	0.77	0.78	0.77	0.78	0.78	0.78	0.79	0.78	0.78
Postoperative	0.65	0.66	0.65	0.68	0.66	0.68	0.68	0.72	0.68	0.72	0.72	0.65	0.70	0.69	0.66
Primary-tumor	0.48	0.54	0.47	0.43	0.46	0.43	0.44	0.53	0.45	0.50	0.50	0.48	0.51	0.54	0.43
Readings-2	1.00	1.00	1.00	1.00	1.00	1.00	1.00	1.00	1.00	1.00	1.00	1.00	1.00	1.00	1.00
Readings-4	1.00	1.00	1.00	1.00	1.00	1.00	1.00	1.00	1.00	1.00	1.00	1.00	1.00	1.00	1.00
Semeion	0.94	0.94	0.94	0.93	0.94	0.93	0.94	0.94	0.94	0.94	0.94	0.97	0.94	0.94	0.94

(continued)

Table 6.2 (continued)

	ACC A	ACC M	ACC H	AUC A	AUC M	AUC H	FM A	FM M	FM H	RAI A	RAI M	RAI H	TPR A	TPR M	TPR H
Shuttle-control	0.63	0.53	0.57	0.52	0.49	0.53	0.58	0.63	0.48	0.58	0.60	0.63	0.56	0.56	0.55
Sick	0.99	0.99	0.99	0.99	0.99	0.98	0.99	0.99	0.99	0.98	0.99	0.99	0.99	0.99	0.98
Solar-flare-1	0.77	0.76	0.76	0.73	0.75	0.73	0.76	0.77	0.76	0.77	0.77	0.76	0.77	0.77	0.76
Solar-flare-2	0.77	0.76	0.76	0.75	0.76	0.75	0.76	0.78	0.76	0.78	0.78	0.77	0.77	0.78	0.76
Sonar	0.88	0.84	0.88	0.85	0.84	0.85	0.88	0.87	0.88	0.88	0.87	0.88	0.88	0.83	0.88
Soybean	0.88	0.85	0.90	0.90	0.87	0.91	0.85	0.89	0.88	0.90	0.87	0.93	0.90	0.87	0.89
Sponge	0.95	0.92	0.94	0.94	0.91	0.92	0.94	0.96	0.94	0.96	0.96	0.95	0.94	0.94	0.93
Tae	0.72	0.75	0.72	0.67	0.64	0.67	0.72	0.76	0.72	0.74	0.77	0.72	0.76	0.75	0.73
Tempdiag	1.00	1.00	1.00	1.00	1.00	1.00	1.00	1.00	1.00	1.00	1.00	1.00	1.00	1.00	1.00
Tep.fea	0.61	0.61	0.61	0.61	0.61	0.61	0.61	0.61	0.61	0.61	0.61	0.61	0.61	0.61	0.61
Tic-tac-toe	0.91	0.95	0.90	0.89	0.95	0.89	0.91	0.94	0.90	0.91	0.94	0.91	0.91	0.96	0.91
Trains	0.87	0.53	0.88	0.78	0.61	0.79	0.77	0.80	0.75	0.88	0.79	0.85	0.66	0.56	0.71
Transfusion	0.79	0.75	0.79	0.78	0.78	0.78	0.79	0.80	0.79	0.80	0.80	0.79	0.80	0.77	0.79
Vehicle	0.84	0.86	0.85	0.80	0.82	0.78	0.85	0.86	0.85	0.86	0.86	0.86	0.87	0.85	0.85
Vote	0.97	0.96	0.97	0.97	0.96	0.96	0.96	0.96	0.96	0.97	0.97	0.97	0.96	0.96	0.97
Wine	0.96	0.94	0.96	0.95	0.94	0.95	0.96	0.95	0.96	0.96	0.95	0.96	0.96	0.93	0.96
Wine-red	0.75	0.79	0.77	0.69	0.73	0.66	0.77	0.80	0.77	0.79	0.80	0.78	0.81	0.80	0.76
Wine-white	0.58	0.61	0.59	0.55	0.58	0.53	0.59	0.62	0.59	0.61	0.61	0.78	0.61	0.62	0.58
Zoo	0.90	0.90	0.84	0.94	0.95	0.94	0.84	0.96	0.84	0.90	0.91	0.90	0.75	0.96	0.81
Average rank	7.94	9.22	8.45	11.30	10.56	12.35	8.17	4.61	9.16	6.27	**3.60**	6.64	5.25	7.17	9.31

Meta-training comprises 5 balanced data sets

Table 6.3 Values are the average performance (rank) of each version of HEAD-DT according to either accuracy or F-Measure

Version	Accuracy rank	F-Measure rank	Average
ACC-A	8.00	7.94	7.97
ACC-M	8.93	9.22	9.08
ACC-H	8.35	8.45	8.40
AUC-A	11.68	11.30	11.49
AUC-M	10.76	10.56	10.66
AUC-H	12.57	12.35	12.46
FM-A	8.25	8.17	8.21
FM-M	4.75	4.61	4.68
FM-H	9.10	9.16	9.13
RAI-A	6.41	6.27	6.34
RAI-M	**3.72**	**3.60**	**3.66**
RAI-H	6.64	6.64	6.64
TPR-A	4.93	5.25	5.09
TPR-M	6.88	7.17	7.03
TPR-H	9.04	9.31	9.18

best-ranked method for either evaluation measure. Only a small position-switching occurs between the accuracy and F-Measure rankings, with respect to the positions of ACC-M, TPR-H, and FM-H.

Table 6.3 summarizes the average rank values obtained by each version of HEAD-DT with respect to accuracy and F-Measure. Values in bold indicate the best performing version according to the corresponding evaluation measure. It can be seen that version RAI-M is the best-performing method regardless of the evaluation measure. The average of the average ranks (average across evaluation measures) indicates the following final ranking positions (from best to worst): (1) RAI-M; (2) FM-M; (3) TPR-A; (4) RAI-A; (5) RAI-H; (6) TPR-M; (7) ACC-A; (8) FM-A; (9) ACC-H; (10) ACC-M; (11) FM-H; (12) TPR-H; (13) AUC-M; (14) AUC-A; (15) AUC-H.

For evaluating whether the differences between versions are statistically significant, we present the critical diagrams of the accuracy and F-Measure values in Fig. 6.1. It is possible to observe that there are no significant differences among the top-4 versions (RAI-M, FM-M, TPR-A, and RAI-A). Nevertheless, RAI-M is the only version that outperforms TPR-M and RAI-H with statistical significance in both evaluation measures, which is not the case of FM-M, TPR-A, and RAI-A.

Some interesting conclusions can be drawn from this first set of experiments with a balanced meta-training set:

• The AUC measure was not particularly effective for evolving decision-tree algorithms in this scenario, regardless of the aggregation scheme being used. Note that versions of HEAD-DT that employ AUC in their fitness function perform quite

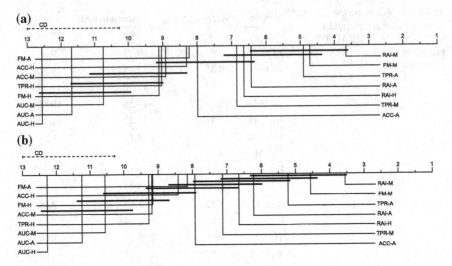

Fig. 6.1 Critical diagrams for the balanced meta-training set experiment. **a** Accuracy rank.
b F-measure rank

poorly when compared to the remaining versions—AUC-M, AUC-A, and AUC-H
are in the bottom of the ranking: 13th, 14th, and 15th position, respectively;
- The use of the harmonic mean as an aggregation scheme was not successful overall. The harmonic mean was often worst aggregation scheme for the evaluation
measures, occupying the lower positions of the ranking (except when combined
to RAI).
- The use of the median, on the other hand, was shown to be very effective in most
cases. For 3 evaluation measures the median was the best aggregation scheme
(relative accuracy improvement, F-Measure, and AUC). In addition, the two best-
ranked versions made use of the median as their aggregation scheme;
- The relative accuracy improvement was overall the best evaluation measure, occupying the top part of the ranking (1st, 4th, and 5th best-ranked versions);
- Finally, both F-Measure and recall were consistently among the best versions (2nd,
3rd, 6th, and 8th best-ranked versions), except once again when associated to the
harmonic mean (11th and 12th).

Figure 6.2 depicts a picture of the fitness evolution throughout the evolutionary
cycle. It presents both the best fitness from the population at a given generation and
the average fitness from the corresponding generation.

Note that version AUC-M (Fig. 6.2e) achieves the perfect fitness from the very
first generation (AUC = 1). We further analysed this particular case and verified that
the decision-tree algorithm designed in this version does not perform any kind of
pruning. Even though prune-free algorithms usually overfit the training data (if no
pre-pruning is performed as well, they achieve 100 % of accuracy in the training data)
and thus underperform in the test data, it seems that this was not the case for the 5
data sets in the meta-training set. In the particular validation sets of the meta-training

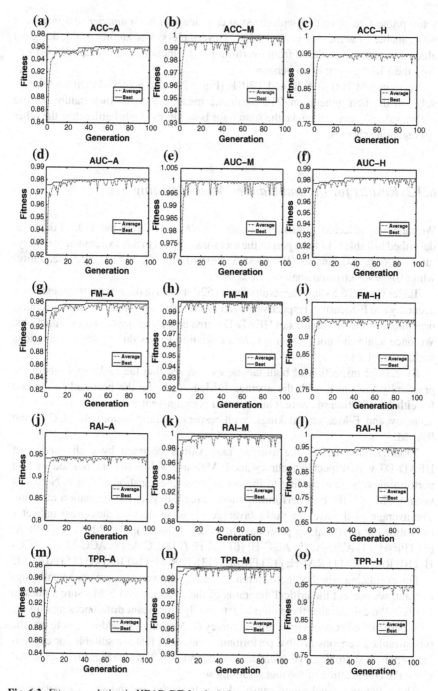

Fig. 6.2 Fitness evolution in HEAD-DT for the balanced meta-training set

set, a prune-free algorithm with the stop criterion *minimum number of 3 instances* was capable of achieving perfect AUC. Nevertheless, this automatically-designed algorithm certainly suffered from overfitting in the meta-test set, since AUC-M was only the 13th-best out of 15 versions.

Versions FM-H (Fig. 6.2i) and TPR-H (Fig. 6.2o) also achieved their best fitness value in the first generation. The harmonic mean, due to its own nature (ignore higher values), seems to make the search for better individuals harder than the other aggregation schemes.

6.3.2 Results for the Imbalanced Meta-Training Set

We randomly selected 5 imbalanced data sets ($IR > 10$) from the 67 UCI data sets described in Table 5.14 to be part of the meta-training set in this experiment: primary-tumor ($IR = 84$), anneal ($IR = 85.5$), arrhythmia ($IR = 122.5$), winequality-white ($IR = 439.6$), and abalone ($IR = 689$).

Tables 6.4 and 6.5 show the results for the 62 data sets in the meta-test set regarding accuracy and F-Measure, respectively. At the bottom of each table, the average rank is presented for the 15 versions of HEAD-DT created by varying the fitness functions. We once again did not present standard deviation values due to space limitations within the tables.

By careful inspection of both tables, we can see that the rankings in them are practically the same, with the average F-Measure being the best-ranked method for either evaluation measure. Only a small position-switching occurs between the accuracy and F-Measure rankings, with respect to the positions of ACC-H and RAI-M.

Table 6.6 summarizes the average rank values obtained by each version of HEAD-DT with respect to accuracy and F-Measure. Values in bold indicate the best performing version according to the corresponding evaluation measure. Note that version FM-A is the best-performing method regardless of the evaluation measure. The average of the average ranks (average across evaluation measures) indicates the following final ranking positions (from best to worst): (1) FM-A; (2) TPR-A; (3) TPR-H; (4) AUC-A; (5) AUC-H; (6) FM-H; (7) ACC-A; (8) ACC-M; (9) ACC-H; (10) RAI-M; (11) RAI-H; (12) FM-M; (13) TPR-M; (14) RAI-A; (15) AUC-M.

For evaluating whether the differences among the versions are statistically significant, we present the critical diagrams of the accuracy and F-Measure values in Fig. 6.3. We can see that there are no statistically significant differences among the 7 (5) best-ranked versions regarding accuracy (F-Measure). In addition, note that the 6 best-ranked versions involve performance measures that are suitable for evaluating imbalanced problems (F-Measure, recall, and AUC), which is actually expected given the composition of the meta-training set.

The following conclusions can be drawn from this second set of experiments concerning imbalanced data sets:

Table 6.4 Accuracy values for the 15 versions of HEAD-DT varying the fitness functions

	ACC			AUC			FM			RAI			TPR		
	A	M	H	A	M	H	A	M	H	A	M	H	A	M	H
Audiology	0.67	0.65	0.69	0.75	0.60	0.60	0.59	0.61	0.60	0.59	0.59	0.55	0.60	0.59	0.60
Autos	0.79	0.74	0.72	0.84	0.63	0.76	0.74	0.49	0.55	0.44	0.47	0.71	0.77	0.35	0.69
Balance-scale	0.82	0.79	0.81	0.80	0.66	0.69	0.72	0.71	0.71	0.58	0.71	0.80	0.71	0.71	0.71
Breast-cancer	0.74	0.73	0.72	0.74	0.60	0.86	0.86	0.70	0.84	0.68	0.84	0.72	0.85	0.62	0.86
Bridges1	0.67	0.69	0.66	0.77	0.53	0.87	0.89	0.82	0.84	0.72	0.78	0.68	0.84	0.75	0.84
Bridges2	0.64	0.67	0.66	0.73	0.53	0.68	0.70	0.58	0.69	0.55	0.60	0.68	0.67	0.56	0.66
Car	0.94	0.92	0.94	0.91	0.76	1.00	1.00	0.94	1.00	0.80	1.00	0.93	1.00	0.94	1.00
Heart-c	0.80	0.80	0.80	0.83	0.67	0.76	0.79	0.75	0.76	0.64	0.76	0.80	0.78	0.74	0.79
cmc	0.60	0.58	0.57	0.57	0.50	0.73	0.74	0.60	0.67	0.59	0.66	0.57	0.74	0.59	0.73
Column-2C	0.85	0.84	0.83	0.86	0.69	0.68	0.74	0.61	0.67	0.60	0.66	0.83	0.73	0.56	0.74
Column-3C	0.84	0.84	0.82	0.87	0.70	0.66	0.71	0.50	0.57	0.54	0.59	0.83	0.68	0.47	0.65
Credit-a	0.87	0.88	0.87	0.88	0.70	0.93	0.95	0.83	0.94	0.75	0.91	0.87	0.95	0.77	0.95
Cylinder-bands	0.78	0.74	0.72	0.78	0.62	0.77	0.81	0.72	0.76	0.64	0.77	0.73	0.80	0.68	0.82
Dermatology	0.96	0.95	0.93	0.96	0.73	0.86	0.88	0.84	0.87	0.69	0.84	0.92	0.88	0.84	0.88
Ecoli	0.84	0.85	0.84	0.86	0.68	0.87	0.88	0.85	0.85	0.69	0.84	0.84	0.88	0.84	0.88
Flags	0.72	0.68	0.64	0.71	0.56	0.96	0.95	0.92	0.94	0.75	0.91	0.66	0.95	0.92	0.94
Credit-g	0.76	0.75	0.75	0.77	0.63	0.74	0.79	0.70	0.78	0.62	0.73	0.75	0.78	0.70	0.77
Glass	0.79	0.79	0.75	0.78	0.62	0.84	0.87	0.77	0.84	0.68	0.80	0.76	0.86	0.71	0.85
Haberman	0.77	0.75	0.75	0.76	0.62	0.85	0.86	0.82	0.86	0.69	0.85	0.76	0.87	0.83	0.87
Hayes-roth	0.85	0.84	0.78	0.85	0.60	0.73	0.77	0.73	0.73	0.61	0.74	0.81	0.76	0.72	0.75
Heart-statlog	0.82	0.81	0.80	0.82	0.67	0.93	0.94	0.92	0.95	0.76	0.94	0.80	0.94	0.92	0.93
Hepatitis	0.81	0.81	0.81	0.89	0.68	0.89	0.86	0.81	0.82	0.66	0.80	0.80	0.86	0.81	0.85

(continued)

Table 6.4 (continued)

	ACC	ACC	ACC	AUC	AUC	AUC	FM	FM	FM	RAI	RAI	RAI	TPR	TPR	TPR
	A	M	H	A	M	H	A	M	H	A	M	H	A	M	H
Colic	0.87	0.85	0.85	0.86	0.67	0.79	0.84	0.79	0.84	0.67	0.80	0.86	0.82	0.79	0.82
Heart-h	0.82	0.81	0.81	0.82	0.66	0.82	0.85	0.82	0.82	0.68	0.82	0.81	0.84	0.83	0.85
Ionosphere	0.93	0.91	0.91	0.92	0.73	0.79	0.79	0.78	0.78	0.62	0.77	0.92	0.81	0.78	0.80
Iris	0.95	0.95	0.95	0.96	0.77	0.82	0.84	0.81	0.82	0.67	0.81	0.95	0.83	0.80	0.84
kr-vs-kp	0.96	0.96	0.96	0.96	0.78	0.73	0.74	0.72	0.72	0.59	0.73	0.96	0.75	0.72	0.74
Labor	0.85	0.77	0.83	0.86	0.68	0.83	0.83	0.79	0.80	0.65	0.80	0.78	0.83	0.79	0.83
Liver-disorders	0.71	0.74	0.71	0.75	0.61	0.83	0.85	0.70	0.83	0.66	0.73	0.73	0.84	0.69	0.80
Lung-cancer	0.62	0.65	0.55	0.65	0.47	0.72	0.74	0.62	0.74	0.58	0.66	0.64	0.74	0.60	0.73
Lymph	0.84	0.80	0.80	0.85	0.65	0.85	0.89	0.78	0.83	0.70	0.83	0.79	0.87	0.67	0.89
Meta.data	0.13	0.12	0.11	0.10	0.13	0.78	0.83	0.77	0.81	0.66	0.79	0.11	0.82	0.78	0.82
Morphological	0.74	0.73	0.73	0.73	0.60	0.95	0.96	0.95	0.96	0.77	0.94	0.73	0.97	0.94	0.96
mb-promoters	0.86	0.85	0.80	0.88	0.63	0.96	0.96	0.95	0.96	0.77	0.95	0.85	0.96	0.96	0.96
Mushroom	0.99	0.98	0.99	0.99	0.80	0.86	0.88	0.85	0.86	0.70	0.86	0.99	0.87	0.82	0.86
Diabetes	0.81	0.78	0.78	0.78	0.64	0.87	0.95	0.86	0.95	0.75	0.87	0.78	0.95	0.80	0.95
Postoperative	0.72	0.71	0.71	0.70	0.57	0.86	0.85	0.79	0.78	0.66	0.78	0.71	0.86	0.77	0.83
Segment	0.96	0.94	0.95	0.94	0.77	0.86	0.90	0.86	0.88	0.71	0.88	0.95	0.90	0.79	0.89
Semeion	0.94	0.92	0.93	0.94	0.76	0.10	0.22	0.09	0.13	0.17	0.13	0.93	0.22	0.08	0.18
Readings-2	0.94	0.98	1.00	1.00	0.80	0.92	0.94	0.90	0.93	0.75	0.92	0.95	0.94	0.88	0.94
Readings-4	0.94	0.98	1.00	1.00	0.80	0.92	0.97	0.91	0.96	0.77	0.95	0.95	0.96	0.79	0.97
Shuttle-control	0.60	0.61	0.61	0.57	0.61	0.87	0.98	0.92	0.97	0.78	0.93	0.60	0.98	0.84	0.98
Sick	0.97	0.95	0.98	0.98	0.79	0.75	0.79	0.69	0.70	0.58	0.59	0.97	0.81	0.59	0.78

(continued)

Table 6.4 (continued)

	ACC	ACC	ACC	AUC	AUC	AUC	FM	FM	FM	RAI	RAI	RAI	TPR	TPR	TPR
	A	M	H	A	M	H	A	M	H	A	M	H	A	M	H
Solar-flare-1	0.72	0.73	0.73	0.74	0.60	0.75	0.77	0.75	0.75	0.61	0.75	0.73	0.76	0.74	0.76
Solar-flare2	0.76	0.75	0.76	0.75	0.61	0.72	0.76	0.72	0.75	0.60	0.72	0.75	0.77	0.72	0.76
Sonar	0.80	0.80	0.79	0.84	0.67	1.00	0.95	1.00	0.94	0.75	0.98	0.80	1.00	1.00	1.00
Soybean	0.79	0.88	0.83	0.87	0.67	0.58	0.63	0.57	0.62	0.49	0.58	0.68	0.62	0.57	0.62
Sponge	0.93	0.93	0.92	0.95	0.74	0.77	0.83	0.75	0.81	0.65	0.74	0.92	0.83	0.75	0.82
kdd-synthetic	0.92	0.92	0.91	0.95	0.74	0.62	0.74	0.62	0.71	0.57	0.64	0.91	0.75	0.60	0.74
Tae	0.66	0.62	0.59	0.70	0.53	0.77	0.83	0.74	0.81	0.65	0.72	0.61	0.82	0.69	0.81
Tempdiag	1.00	1.00	0.91	1.00	0.77	0.77	0.81	0.75	0.78	0.64	0.76	0.95	0.81	0.75	0.81
Tep.fea	0.65	0.65	0.65	0.65	0.52	0.81	0.80	0.79	0.82	0.63	0.81	0.65	0.93	0.58	0.90
Tic-tac-toe	0.90	0.88	0.88	0.90	0.73	1.00	0.95	1.00	0.94	0.75	0.98	0.90	1.00	1.00	1.00
Trains	0.59	0.48	0.37	0.75	0.39	0.65	0.65	0.65	0.65	0.52	0.65	0.51	0.65	0.65	0.65
Transfusion	0.77	0.77	0.79	0.79	0.64	0.95	0.94	0.90	0.93	0.74	0.92	0.78	0.95	0.90	0.94
Vehicle	0.79	0.76	0.74	0.76	0.64	0.94	0.96	0.93	0.95	0.76	0.93	0.74	0.96	0.94	0.96
Vote	0.96	0.96	0.96	0.96	0.77	0.98	0.98	0.97	0.98	0.78	0.96	0.96	0.99	0.98	0.98
Vowel	0.73	0.75	0.74	0.67	0.62	0.96	0.96	0.95	0.96	0.76	0.95	0.70	0.95	0.95	0.95
Wine	0.93	0.90	0.90	0.96	0.75	0.94	0.99	0.95	0.98	0.78	0.94	0.91	0.99	0.91	0.99
Wine-red	0.69	0.65	0.63	0.62	0.55	0.99	1.00	0.99	1.00	0.80	0.98	0.63	1.00	0.98	1.00
Breast-w	0.95	0.94	0.95	0.95	0.76	0.94	0.97	0.95	0.97	0.78	0.95	0.95	0.98	0.95	0.97
Zoo	0.94	0.92	0.89	0.93	0.65	0.66	0.82	0.65	0.79	0.61	0.70	0.92	0.88	0.64	0.87
Average rank	6.70	7.94	8.40	5.87	13.43	6.70	**4.02**	9.71	6.70	13.40	8.58	8.72	4.19	10.53	5.10

Meta-training comprises 5 imbalanced data sets

Table 6.5 F-Measure values for the 15 versions of HEAD-DT varying the fitness functions

	ACC A	ACC M	ACC H	AUC A	AUC M	AUC H	FM A	FM M	FM H	RAI A	RAI M	RAI H	TPR A	TPR M	TPR H
Audiology	0.63	0.60	0.64	0.71	0.56	0.58	0.54	0.51	0.52	0.52	0.50	0.48	0.48	0.46	0.47
Autos	0.79	0.74	0.71	0.84	0.63	0.76	0.73	0.46	0.54	0.43	0.45	0.71	0.76	0.30	0.68
Balance-scale	0.81	0.76	0.77	0.78	0.65	0.66	0.67	0.59	0.63	0.53	0.61	0.77	0.60	0.59	0.61
Breast-cancer	0.68	0.67	0.68	0.73	0.59	0.86	0.86	0.67	0.84	0.68	0.84	0.65	0.85	0.62	0.86
Bridges1	0.62	0.63	0.60	0.77	0.51	0.87	0.89	0.81	0.82	0.71	0.75	0.63	0.83	0.70	0.83
Bridges2	0.57	0.60	0.59	0.72	0.50	0.68	0.70	0.58	0.69	0.55	0.60	0.61	0.67	0.56	0.66
Car	0.93	0.92	0.94	0.91	0.76	1.00	1.00	0.94	1.00	0.80	1.00	0.93	1.00	0.94	1.00
Heart-c	0.80	0.80	0.80	0.83	0.67	0.74	0.78	0.69	0.70	0.62	0.71	0.80	0.76	0.69	0.77
cmc	0.59	0.57	0.57	0.57	0.49	0.73	0.73	0.53	0.61	0.58	0.59	0.57	0.72	0.53	0.71
Column-2C	0.85	0.84	0.83	0.86	0.69	0.67	0.73	0.52	0.62	0.58	0.59	0.83	0.71	0.48	0.72
Column-3C	0.83	0.84	0.82	0.87	0.70	0.66	0.71	0.40	0.53	0.54	0.57	0.82	0.68	0.40	0.65
Credit-a	0.87	0.88	0.87	0.88	0.70	0.93	0.95	0.78	0.94	0.75	0.89	0.87	0.95	0.73	0.95
Cylinder-bands	0.78	0.73	0.72	0.78	0.62	0.77	0.81	0.70	0.75	0.63	0.76	0.73	0.80	0.67	0.81
Dermatology	0.96	0.95	0.93	0.96	0.73	0.86	0.88	0.84	0.87	0.69	0.84	0.91	0.88	0.84	0.87
Ecoli	0.83	0.84	0.83	0.85	0.68	0.87	0.88	0.85	0.84	0.69	0.84	0.83	0.88	0.83	0.88
Flags	0.70	0.67	0.62	0.71	0.56	0.96	0.95	0.92	0.94	0.75	0.91	0.63	0.95	0.92	0.94
Credit-g	0.72	0.74	0.72	0.76	0.62	0.74	0.79	0.69	0.78	0.62	0.72	0.73	0.78	0.69	0.77
Glass	0.78	0.78	0.74	0.77	0.61	0.84	0.87	0.75	0.84	0.68	0.79	0.74	0.85	0.70	0.85
Haberman	0.71	0.71	0.70	0.75	0.61	0.84	0.86	0.80	0.86	0.68	0.84	0.70	0.86	0.81	0.86
Hayes-roth	0.85	0.83	0.78	0.85	0.59	0.72	0.76	0.68	0.68	0.60	0.70	0.81	0.74	0.68	0.73
Heart-statlog	0.82	0.81	0.80	0.82	0.67	0.93	0.93	0.88	0.94	0.75	0.92	0.80	0.93	0.89	0.89
Hepatitis	0.76	0.75	0.76	0.88	0.66	0.88	0.85	0.76	0.77	0.64	0.76	0.74	0.84	0.76	0.83

(continued)

Table 6.5 (continued)

	ACC	ACC	ACC	AUC	AUC	AUC	FM	FM	FM	RAI	RAI	RAI	TPR	TPR	TPR
	A	M	H	A	M	H	A	M	H	A	M	H	A	M	H
Colic	0.87	0.85	0.85	0.86	0.66	0.76	0.83	0.77	0.83	0.67	0.77	0.86	0.80	0.77	0.80
Heart-h	0.82	0.80	0.80	0.82	0.66	0.82	0.85	0.82	0.82	0.68	0.82	0.81	0.84	0.83	0.84
Ionosphere	0.93	0.91	0.91	0.92	0.73	0.77	0.78	0.73	0.74	0.62	0.73	0.92	0.78	0.76	0.78
Iris	0.95	0.95	0.95	0.96	0.77	0.82	0.84	0.80	0.82	0.67	0.81	0.95	0.83	0.80	0.84
kr-vs-kp	0.96	0.96	0.96	0.96	0.78	0.72	0.73	0.69	0.70	0.58	0.70	0.96	0.73	0.70	0.72
Labor	0.83	0.72	0.83	0.85	0.68	0.83	0.83	0.79	0.80	0.65	0.79	0.74	0.83	0.79	0.83
Liver-disorders	0.69	0.73	0.70	0.75	0.61	0.83	0.85	0.69	0.83	0.66	0.73	0.72	0.84	0.68	0.80
Lung-cancer	0.60	0.64	0.48	0.65	0.42	0.71	0.74	0.59	0.73	0.57	0.64	0.63	0.73	0.56	0.72
Lymph	0.83	0.80	0.78	0.85	0.65	0.85	0.89	0.77	0.83	0.70	0.83	0.78	0.87	0.67	0.89
Meta.data	0.11	0.10	0.09	0.09	0.12	0.78	0.83	0.77	0.81	0.65	0.78	0.08	0.82	0.77	0.82
Morphological	0.72	0.71	0.72	0.72	0.60	0.95	0.96	0.95	0.96	0.77	0.94	0.72	0.97	0.94	0.96
mb-promoters	0.86	0.84	0.80	0.88	0.63	0.96	0.96	0.95	0.96	0.77	0.95	0.85	0.96	0.96	0.96
Mushroom	0.99	0.98	0.99	0.99	0.80	0.86	0.88	0.85	0.86	0.70	0.86	0.99	0.87	0.82	0.86
Diabetes	0.80	0.78	0.77	0.78	0.64	0.87	0.95	0.86	0.94	0.75	0.86	0.77	0.95	0.80	0.95
Postoperative	0.63	0.59	0.59	0.65	0.51	0.86	0.85	0.79	0.78	0.66	0.78	0.59	0.86	0.77	0.83
Segment	0.96	0.94	0.95	0.94	0.77	0.85	0.90	0.86	0.88	0.71	0.88	0.95	0.90	0.78	0.89
Semeion	0.94	0.90	0.92	0.93	0.76	0.09	0.21	0.07	0.11	0.16	0.11	0.92	0.21	0.05	0.17
Readings-2	0.93	0.98	1.00	1.00	0.80	0.92	0.94	0.89	0.93	0.75	0.92	0.94	0.94	0.88	0.94
Readings-4	0.93	0.98	1.00	1.00	0.80	0.92	0.97	0.91	0.96	0.77	0.95	0.94	0.96	0.79	0.97
Shuttle-control	0.52	0.49	0.47	0.55	0.52	0.87	0.98	0.92	0.97	0.78	0.92	0.47	0.98	0.83	0.98
Sick	0.97	0.94	0.98	0.98	0.79	0.71	0.77	0.63	0.68	0.56	0.54	0.96	0.79	0.51	0.75

(continued)

Table 6.5 (continued)

	ACC			AUC			FM			RAI			TPR		
	A	M	H	A	M	H	A	M	H	A	M	H	A	M	H
Solar-flare-1	0.71	0.71	0.71	0.72	0.59	0.74	0.76	0.73	0.73	0.60	0.73	0.70	0.75	0.72	0.74
Solar-flare2	0.74	0.73	0.73	0.74	0.60	0.72	0.76	0.70	0.74	0.60	0.72	0.73	0.77	0.69	0.76
Sonar	0.80	0.80	0.79	0.84	0.67	1.00	0.93	1.00	0.93	0.74	0.97	0.80	1.00	1.00	1.00
Soybean	0.77	0.87	0.81	0.86	0.66	0.57	0.63	0.57	0.62	0.49	0.58	0.65	0.62	0.57	0.61
Sponge	0.91	0.91	0.88	0.94	0.73	0.76	0.83	0.74	0.81	0.65	0.74	0.88	0.83	0.74	0.82
kdd-synthetic	0.92	0.92	0.91	0.95	0.74	0.61	0.74	0.60	0.71	0.56	0.63	0.91	0.75	0.57	0.73
Tae	0.66	0.61	0.59	0.70	0.53	0.76	0.83	0.73	0.81	0.65	0.72	0.61	0.82	0.69	0.80
Tempdiag	1.00	1.00	0.91	1.00	0.77	0.76	0.81	0.71	0.75	0.64	0.75	0.95	0.81	0.73	0.80
Tep.fea	0.61	0.61	0.60	0.61	0.49	0.79	0.78	0.77	0.80	0.61	0.78	0.61	0.93	0.51	0.89
Tic-tac-toe	0.90	0.88	0.88	0.91	0.73	1.00	0.93	1.00	0.93	0.74	0.97	0.90	1.00	1.00	1.00
Trains	0.59	0.47	0.33	0.75	0.37	0.61	0.61	0.61	0.61	0.49	0.61	0.49	0.61	0.61	0.61
Transfusion	0.73	0.71	0.77	0.77	0.63	0.96	0.94	0.90	0.93	0.74	0.92	0.73	0.95	0.90	0.94
Vehicle	0.79	0.75	0.74	0.76	0.64	0.93	0.96	0.92	0.95	0.76	0.92	0.73	0.96	0.94	0.96
Vote	0.96	0.96	0.96	0.96	0.77	0.98	0.98	0.97	0.98	0.78	0.96	0.96	0.99	0.98	0.98
Vowel	0.73	0.75	0.74	0.66	0.62	0.96	0.96	0.95	0.96	0.76	0.95	0.69	0.95	0.95	0.95
Wine	0.93	0.90	0.90	0.96	0.75	0.94	0.99	0.95	0.98	0.78	0.94	0.91	0.99	0.91	0.99
Wine-red	0.68	0.63	0.61	0.61	0.54	0.99	1.00	0.99	1.00	0.80	0.98	0.61	1.00	0.98	1.00
Breast-w	0.95	0.94	0.95	0.95	0.76	0.94	0.97	0.95	0.97	0.78	0.95	0.95	0.98	0.95	0.97
Zoo	0.94	0.91	0.86	0.93	0.62	0.65	0.82	0.65	0.79	0.61	0.69	0.91	0.88	0.64	0.87
Average rank	6.92	8.23	8.74	5.44	13.23	6.25	**3.83**	9.97	6.79	12.94	8.65	8.95	4.27	10.56	5.25

Meta-training comprises 5 imbalanced data sets

Table 6.6 Values are the average performance (rank) of each version of HEAD-DT according to either accuracy or F-Measure

Version	Accuracy rank	F-Measure rank	Average
ACC-A	6.70	6.92	6.81
ACC-M	7.94	8.23	8.09
ACC-H	8.40	8.74	8.57
AUC-A	5.87	5.44	5.66
AUC-M	13.43	13.23	13.33
AUC-H	6.70	6.25	6.48
FM-A	**4.02**	**3.83**	**3.93**
FM-M	9.71	9.97	9.84
FM-H	6.70	6.79	6.75
RAI-A	13.40	12.94	13.17
RAI-M	8.58	8.65	8.62
RAI-H	8.72	8.95	8.84
TPR-A	4.19	4.27	4.23
TPR-M	10.53	10.56	10.55
TPR-H	5.10	5.25	5.18

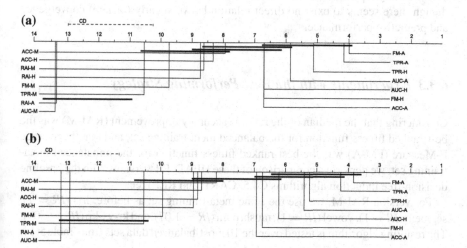

Fig. 6.3 Critical diagrams for the imbalanced meta-training set experiment. **a** Accuracy rank. **b** F-measure rank

- The relative accuracy improvement is not suitable for dealing with imbalanced data sets and hence occupies the bottom positions of the ranking (10th, 11th, and 14th positions). This behavior is expected given that RAI measures the improvement over the majority-class accuracy, and such an improvement is often damaging for imbalanced problems, in which the goal is to improve the accuracy of the less-frequent class(es);
- The median was the worst aggregation scheme overall, figuring in the bottom positions of the ranking (8th, 10th, 12th, 13th, and 15th). It is interesting to notice that the median was very successful in the balanced meta-training experiment, and quite the opposite in the imbalanced one;
- The simple average, on the other hand, presented itself as the best aggregation scheme for the imbalanced data, figuring in the top of the ranking (1st, 2nd, 4th, 7th), except when associated to RAI (14th), which was the worst performance measure overall;
- The 6 best-ranked versions were those employing performance measures known to be suitable for imbalanced data (F-Measure, recall, and AUC);
- Finally, the harmonic mean had a solid performance throughout this experiment, differently from its performance in the balanced meta-training experiment.

Figure 6.4 depicts a picture of the fitness evolution throughout the evolutionary cycle. Note that whereas some versions find their best individual at the very end of evolution (e.g., FM-H, Fig. 6.4i), others converge quite early (e.g., TPR-H, Fig. 6.4o), though there seems to exist no direct relation between early (or late) convergence and predictive performance.

6.3.3 Experiments with the Best-Performing Strategy

Considering that the median of the relative accuracy improvement (RAI-M) was the best-ranked fitness function for the balanced meta-training set, and that the average F-Measure (FM-A) was the best-ranked fitness function for the imbalanced meta-training set, we perform a comparison of these HEAD-DT versions with the baseline decision-tree induction algorithms C4.5, CART, and REPTree.

For version RAI-M, we use the same meta-training set as before: iris ($IR = 1$), segment ($IR = 1$), vowel ($IR = 1$), mushroom ($IR = 1.07$), and kr-vs-kp ($IR = 1.09$). The resulting algorithm is tested over the 10 most-balanced data sets from Table 5.14:

1. meta-data ($IR = 1$);
2. mfeat ($IR = 1$);
3. mb-promoters ($IR = 1$);
4. kdd-synthetic ($IR = 1$);
5. trains ($IR = 1$);
6. tae ($IR = 1.06$);
7. vehicle ($IR = 1.10$);

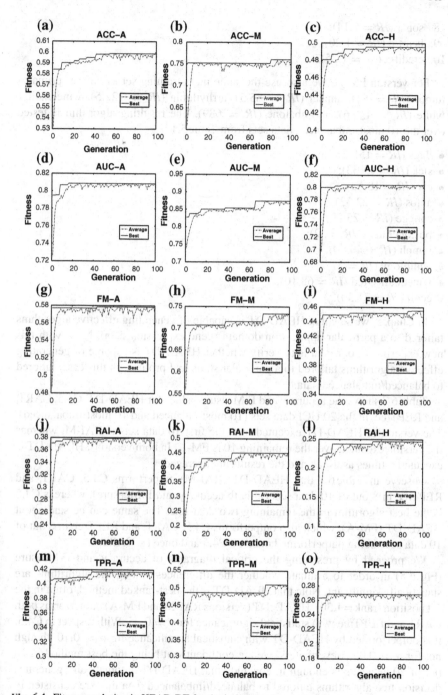

Fig. 6.4 Fitness evolution in HEAD-DT for the imbalanced meta-training set

8. sonar ($IR = 1.14$);
9. heart-c ($IR = 1.20$);
10. credit-a ($IR = 1.25$).

For version FM-A, we also use the same meta-training set as before: primary-tumor ($IR = 84$), anneal ($IR = 85.5$), arrhythmia ($IR = 122.5$), winequality-white ($IR = 439.6$), and abalone ($IR = 689$). The resulting algorithm is tested over the 10 most-imbalanced data sets from Table 5.14:

- flags ($IR = 15$);
- sick ($IR = 15.33$);
- car ($IR = 18.62$);
- autos ($IR = 22.33$);
- sponge ($IR = 23.33$);
- postoperative ($IR = 32$);
- lymph ($IR = 40.50$);
- audiology ($IR = 57$);
- winequality-red ($IR = 68.10$);
- ecoli ($IR = 71.50$).

In Chap. 5, we saw that HEAD-DT is capable of generating effective algorithms tailored to a particular application domain (gene expression data). Now, with this new experiment, our goal is to verify whether HEAD-DT is capable of generating effective algorithms tailored to a particular statistical profile—in this case, tailored to balanced/imbalanced data.

Table 6.7 shows the accuracy and F-Measure values for HEAD-DT, C4.5, CART, and REPTree, in the 20 UCI data sets (10 most-balanced and 10 most-imbalanced). The version of HEAD-DT executed over the first 10 data sets is RAI-M, whereas the version executed over the remaining 10 is FM-A. In both versions, HEAD-DT is executed 5 times as usual, and the results are averaged.

Observe in Table 6.7 that HEAD-DT (RAI-M) outperforms C4.5, CART, and REPTree in 8 out of 10 data sets (in both accuracy and F-Measure), whereas C4.5 is the best algorithm in the remaining two data sets. The same can be said about HEAD-DT (FM-A), which also outperforms C4.5, CART, and REPTree in 8 out of 10 data sets, being outperformed once by C4.5 and once by CART.

We proceed by presenting the critical diagrams of accuracy and F-Measure (Fig. 6.5) in order to evaluate whether the differences among the algorithms are statistically significant. Note that HEAD-DT is the best-ranked method, often in the 1st position (rank = 1.30). HEAD-DT (versions RAI-M and FM-A) outperforms both CART and REPTree with statistical significance for $\alpha = 0.05$. With respect to C4.5, it is outperformed by HEAD-DT with statistical significance for $\alpha = 0.10$, though not for $\alpha = 0.05$. Nevertheless, we are confident that being the best method in 16 out of 20 data sets is enough to conclude that HEAD-DT automatically generates decision-tree algorithms tailored to balanced/imbalanced data that are consistently more effective than C4.5, CART, and REPTree.

Table 6.7 Accuracy and F-Measure values for the 10 most-balanced data sets and the 10 most-imbalanced data sets

Data set	IR	HEAD-DT		C4.5		CART		REP	
		Accuracy	F-Measure	Accuracy	F-Measure	Accuracy	F-Measure	Accuracy	F-Measure
Meta.data	1.00	**0.35 ± 0.09**	**0.35 ± 0.10**	0.04 ± 0.03	0.02 ± 0.02	0.05 ± 0.03	0.02 ± 0.01	0.04 ± 0.00	0.00 ± 0.00
mfeat	1.00	**0.79 ± 0.01**	**0.78 ± 0.02**	0.72 ± 0.02	0.70 ± 0.02	0.72 ± 0.04	0.70 ± 0.04	0.72 ± 0.03	0.70 ± 0.03
mb-promoters	1.00	**0.89 ± 0.03**	**0.89 ± 0.03**	0.80 ± 0.13	0.79 ± 0.14	0.72 ± 0.14	0.71 ± 0.14	0.77 ± 0.15	0.76 ± 0.15
kdd-synthetic	1.00	0.91 ± 0.03	0.91 ± 0.03	**0.91 ± 0.04**	**0.91 ± 0.04**	0.88 ± 0.04	0.88 ± 0.04	0.88 ± 0.03	0.87 ± 0.04
Trains	1.00	0.79 ± 0.06	0.79 ± 0.06	**0.90 ± 0.32**	**0.90 ± 0.32**	0.20 ± 0.42	0.20 ± 0.42	0.00 ± 0.00	0.00 ± 0.00
Tae	1.06	**0.77 ± 0.03**	**0.77 ± 0.03**	0.60 ± 0.11	0.59 ± 0.12	0.51 ± 0.12	0.49 ± 0.15	0.47 ± 0.12	0.45 ± 0.12
Vehicle	1.10	**0.86 ± 0.01**	**0.86 ± 0.01**	0.74 ± 0.04	0.73 ± 0.04	0.72 ± 0.04	0.71 ± 0.05	0.71 ± 0.04	0.70 ± 0.04
Sonar	1.14	**0.87 ± 0.01**	**0.87 ± 0.01**	0.73 ± 0.08	0.72 ± 0.08	0.71 ± 0.06	0.71 ± 0.06	0.71 ± 0.07	0.70 ± 0.07
Heart-c	1.20	**0.87 ± 0.01**	**0.87 ± 0.01**	0.77 ± 0.09	0.76 ± 0.09	0.81 ± 0.04	0.80 ± 0.04	0.77 ± 0.08	0.77 ± 0.08
Credit-a	1.25	**0.90 ± 0.01**	**0.90 ± 0.01**	0.85 ± 0.03	0.85 ± 0.03	0.84 ± 0.03	0.84 ± 0.03	0.85 ± 0.03	0.85 ± 0.03
Flags	15.00	**0.74 ± 0.01**	**0.74 ± 0.01**	0.63 ± 0.05	0.61 ± 0.05	0.61 ± 0.10	0.57 ± 0.10	0.62 ± 0.10	0.58 ± 0.10
Sick	15.33	0.98 ± 0.01	0.98 ± 0.01	0.99 ± 0.00	0.99 ± 0.00	**0.99 ± 0.01**	**0.99 ± 0.01**	0.99 ± 0.01	0.99 ± 0.01
Car	18.62	**0.98 ± 0.01**	**0.98 ± 0.01**	0.93 ± 0.02	0.93 ± 0.02	0.97 ± 0.02	0.97 ± 0.02	0.89 ± 0.02	0.89 ± 0.02
Autos	22.33	0.85 ± 0.03	0.85 ± 0.03	**0.86 ± 0.06**	**0.85 ± 0.07**	0.78 ± 0.10	0.77 ± 0.10	0.65 ± 0.08	0.62 ± 0.07
Sponge	23.33	**0.94 ± 0.01**	**0.93 ± 0.02**	0.93 ± 0.06	0.89 ± 0.09	0.91 ± 0.06	0.88 ± 0.09	0.91 ± 0.08	0.88 ± 0.10
Postoperative	32.00	**0.72 ± 0.01**	**0.67 ± 0.04**	0.70 ± 0.05	0.59 ± 0.07	0.71 ± 0.06	0.59 ± 0.08	0.69 ± 0.09	0.58 ± 0.09
Lymph	40.50	**0.87 ± 0.01**	**0.87 ± 0.01**	0.78 ± 0.09	0.79 ± 0.10	0.75 ± 0.12	0.73 ± 0.14	0.77 ± 0.11	0.76 ± 0.12
Audiology	57.00	**0.79 ± 0.04**	**0.77 ± 0.05**	0.78 ± 0.07	0.75 ± 0.08	0.74 ± 0.05	0.71 ± 0.05	0.74 ± 0.08	0.70 ± 0.09
Wine-red	68.10	**0.74 ± 0.02**	**0.74 ± 0.02**	0.61 ± 0.03	0.61 ± 0.03	0.63 ± 0.02	0.61 ± 0.02	0.60 ± 0.03	0.58 ± 0.03
Ecoli	71.50	**0.86 ± 0.01**	**0.86 ± 0.01**	0.84 ± 0.07	0.83 ± 0.07	0.84 ± 0.07	0.82 ± 0.07	0.79 ± 0.09	0.77 ± 0.09
Rank		**1.30**	**1.30**	2.25	2.25	2.90	2.90	3.55	3.55

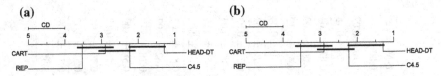

Fig. 6.5 Critical diagrams for accuracy and F-Measure. Values are regarding the 20 UCI data sets in Table 6.7. **a** Accuracy rank for the balanced data sets. **b** F-measure rank for the balanced data sets

Since HEAD-DT is run 5 times for alleviating the randomness effect of evolutionary algorithms, we further analyse the 5 algorithms generated by HEAD-DT for the balanced meta-training set and the 5 algorithms generated for the imbalanced meta-training set.

Regarding the balanced meta-training set, we noticed that the favored split criterion was the G statistic (present in 40 % of the algorithms). The favored stop criterion was stopping the tree-splitting process only when there is a single instance in the node (present in 80 % of the algorithms). The homogeneous stop was present in the remaining 20 % of the algorithms, but since a single instance is always homogeneous (only 1 class represented in the node), we can say that HEAD-DT stop criterion was actually stop splitting nodes when they are homogeneous. Surprisingly, the favored pruning strategy was not to use any pruning strategy (80 % of the algorithms). It seems that this particular combination of design components did not lead to overfitting, even though the trees were not pruned at any point. Algorithm 1 shows this custom algorithm designed for balanced data sets.

Algorithm 1 Custom algorithm designed by HEAD-DT (RAI-M) for balanced data sets.

1: Recursively split nodes using the G statistic;
2: Perform nominal splits in multiple subsets;
3: Perform step 1 until class-homogeneity;
4: Do not perform any pruning strategy;
         When dealing with missing values:
5: Calculate the split of missing values by weighting the split criterion value;
6: Distribute missing values by weighting them according to partition probability;
7: For classifying an instance with missing values, halt in the current node.

Regarding the imbalanced meta-training set, we noticed that two split criteria stand out: DCSM (present in 40 % of the algorithms) and Normalized Gain (also present in 40 % of the algorithms). In 100 % of the algorithms, the nominal splits were aggregated into binary splits. The favored stop criterion was either the homogeneous stop (60 % of the algorithms) of the algorithms or tree stop when a maximum depth of around 10 levels is reached (40 % of the algorithms). Finally, the pruning strategy was also divided between PEP pruning with 1 SE (40 % of the algorithms) and no pruning at all (40 % of the algorithms). We noticed that whenever the algorithm employed DCSM, PEP pruning was the favored pruning strategy. Similarly, whenever the Normalized Gain was selected, *no pruning* was the favored pruning strategy. It seems that HEAD-DT was capable of detecting a correlation between different split criteria and pruning strategies. Algorithm 2 shows the custom algorithm that was tailored to imbalanced data (we actually present the choices of different components when it was the case).

Algorithm 2 Custom algorithm designed by HEAD-DT (FM-A) for imbalanced data sets.

1: Recursively split nodes using either DCSM or the Normalized Gain;
2: Aggregate nominal splits into binary subsets;
3: Perform step 1 until class-homogeneity or a maximum depth of 9 (10) levels;
4: Either do not perform pruning and remove nodes that do not reduce training error, or perform PEP pruning with 1 SE;
    When dealing with missing values:
5: Ignore missing values or perform unsupervised imputation when calculating the split criterion;
6: Perform unsupervised imputation before distributing missing values;
7: For classifying an instance with missing values, halt in the current node or explore all branches and combine the classification.

Regarding the missing value strategies, we did not notice any particular pattern in either the balanced or the imbalanced scenarios. Hence, the missing-value strategies presented in Algorithms 1 and 2 are only examples of selected components, though they did not stand out in terms of appearance frequency.

6.4 Chapter Remarks

In this chapter, we performed a series of experiments to analyse in more detail the impact of different fitness functions during the evolutionary cycle of HEAD-DT. In the first part of the chapter, we presented 5 classification performance measures and three aggregation schemes to combine these measures during fitness evaluation of multiple data sets. The combination of performance measures and aggregation schemes resulted in 15 different versions of HEAD-DT.

We designed two experimental scenarios to evaluate the 15 versions of HEAD-DT. In the first scenario, HEAD-DT is executed on a meta-training set with 5 balanced data sets, and on a meta-test set with the remaining 62 available UCI data sets. In the second scenario, HEAD-DT is executed on a meta-training set with 5 imbalanced data sets, and the meta-test set with the remaining 62 available UCI data sets. For measuring the level of data set balance, we make use of the imbalance ratio (IR), which is the ratio between the most-frequent and the less-frequent classes of the data.

Results of the experiments indicated that the median of the relative accuracy improvement was the most suitable fitness function for the balanced scenario, whereas the average of the F-Measure was the most suitable fitness function for the imbalanced scenario. The next step of the empirical analysis was to compare these versions of HEAD-DT with the baseline decision-tree induction algorithms C4.5, CART, and REPTree. For such, we employed the same meta-training sets than before, though the meta-test sets exclusively comprised balanced (imbalanced) data sets. The experimental results confirmed that HEAD-DT can generate algorithms tailored to a particular statistical profile (data set balance) that are more effective than C4.5, CART, and REPTree, outperforming them in 16 out of 20 data sets.

References

1. T. Fawcett, An introduction to ROC analysis. Pattern Recognit. Lett. **27**(8), 861–874 (2006)
2. C. Ferri, J. Hernández-Orallo, R. Modroiu, An experimental comparison of performance measures for classification. Pattern Recognit. Lett. **30**(1), 27–38 (2009)
3. B. Hanczar et al., Small-sample precision of ROC-related estimates. Bioinformatics **26**(6), 822–830 (2010)
4. D.J. Hand, Measuring classifier performance: a coherent alternative to the area under the ROC curve. Mach. Learn. **77**(1), 103–123 (2009)
5. J.M. Lobo, A. Jiménez-Valverde, R. Real, AUC: a misleading measure of the performance of predictive distribution models. Glob. Ecol. Biogeogr. **17**(2), 145–151 (2008)
6. S.J. Mason, N.E. Graham, Areas beneath the relative operating characteristics (roc) and relative operating levels (rol) curves: statistical significance and interpretation. Q. J. R. Meteorol. Soc. **128**(584), 2145–2166 (2002)
7. G.L. Pappa, Automatically evolving rule induction algorithms with grammar-based genetic programming, Ph.D. thesis. University of Kent at Canterbury (2007)
8. D. Powers, Evaluation: From precision, recall and f-measure to ROC, informedness, markedness and correlation. J. Mach. Learn. Technol. **2**(1), 37–63 (2011)

Chapter 7
Conclusions

We presented in this book an approach for the automatic design of decision-tree induction algorithms, namely HEAD-DT (Hyper-Heuristic Evolutionary Algorithm for Automatically Designing Decision-Tree Induction Algorithms). HEAD-DT makes use of an evolutionary algorithm to perform a search in the space of more than 21 million decision-tree induction algorithms. The search is guided by the performance of the candidate algorithms in a *meta-training set*, and it may follow two distinct frameworks:

- Evolution of a decision-tree induction algorithm tailored to one specific data set at a time (specific framework);
- Evolution of a decision-tree induction algorithm from multiple data sets (general framework).

We carried out extensive experimentation with both specific and general frameworks. In the first, HEAD-DT uses data from a single data set in both meta-training and meta-test sets. The goal is to design a given decision-tree induction algorithm so it excels at that specific data set, and no requirements are made regarding its performance in other data sets. Experiments with 20 UCI data sets showed that HEAD-DT significantly outperforms algorithms like CART [5] and C4.5 [11] with respect to performance measures such as accuracy and F-Measure.

In the second framework, HEAD-DT was further tested with three distinct objectives:

1. To evolve a single decision-tree algorithm tailor-made to data sets from a particular application domain (homogeneous approach).
2. To evolve a single decision-tree algorithm robust across a variety of different data sets (heterogeneous approach).
3. To evolve a single decision-tree algorithm tailored to data sets that share a particular statistical profile.

In order to evaluate objective 1, we performed a thorough empirical analysis on 35 microarray gene expression data sets [12]. The experimental results indicated that automatically-designed decision-tree induction algorithms tailored to a particular domain (in this case, microarray data) usually outperform traditional decision-tree algorithms like C4.5 and CART.

© The Author(s) 2015
R.C. Barros et al., *Automatic Design of Decision-Tree Induction Algorithms*,
SpringerBriefs in Computer Science, DOI 10.1007/978-3-319-14231-9_7

For evaluating objective 2, we conducted a thorough empirical analysis on 67 UCI public data sets [7]. According to the experimental results, the automatically-designed "all-around" decision-tree induction algorithms, which are meant to be robust across very different data sets, presented a performance similar to traditional decision-tree algorithms like C4.5 and CART, even though they seemed to be suffering from *meta-overfitting*.

With respect to objective 3, we first performed an extensive analysis with 15 distinct fitness functions for HEAD-DT. We concluded that the best versions of HEAD-DT were able to automatically design decision-tree induction algorithms tailored to balanced (and imbalanced) data sets which consistently outperformed traditional algorithms like C4.5 and CART.

Overall, HEAD-DT presented a good performance in the four different investigated scenarios (one scenario regarding the specific framework and three scenarios regarding the general framework). Next, we present the limitations (Sect. 7.1) and future work possibilities (Sect. 7.2) we envision for continuing the study presented in this book.

7.1 Limitations

HEAD-DT has the intrinsic disadvantage of evolutionary algorithms, which is a high execution time. Furthermore, it inherits the disadvantages of a hyper-heuristic approach, which is the need of evaluating several data sets in the meta-training set (at least in the general framework), also translating into additional execution time. However, we recall from Chap. 5, Sect. 5.4, that HEAD-DT may be seen as a fast method for automatically designing decision-tree algorithms. Even though it may take up to several hours to generate a tailor-made algorithm for a given application domain, its further application in data sets from the domain of interest is just as fast as most traditional top-down decision-tree induction algorithms. The cost of developing a new decision-tree algorithm from scratch to a particular domain, on the other hand, would be in the order of several months.

We believe that the main limitation in the current implementation of HEAD-DT is the meta-training set setup. Currently, we employed a random methodology to select data sets to be part of the meta-training set in the general framework. Even though randomly-generated meta-training sets still provided good results for the homogeneous approach in the general framework, we believe that automatic, intelligent construction of meta-training sets can significantly improve HEAD-DT's predictive performance. Some suggestions regarding this matter are discussed in the next section.

Finally, the *meta-overfitting* problem identified in Chap. 5, Sect. 5.2.3, is also a current limitation of the HEAD-DT implementation. Probably the easiest way of solving this problem would be to feed the meta-training set with a large variety of data sets. However, this solution would slow down evolution significantly, taking HEAD-DT's average execution time from a few hours to days or weeks. We comment in the next section some other alternatives to the *meta-overfitting* problem.

7.2 Opportunities for Future Work

In this section, we discuss seven new research opportunities resulting from this study, namely: (i) proposing an extended HEAD-DT's genome that takes into account new induction strategies, oblique splits, and the ability to deal with regression problems; (ii) proposing a new multi-objective fitness function; (iii) proposing a new method for automatically selecting proper data sets to be part of the meta-training set; (iv) proposing a parameter-free evolutionary search; (v) proposing solutions to the meta-overfitting problem; (vi) proposing the evolution of decision-tree induction algorithms to compose an ensemble; and (vii) proposing a new genetic search that makes use of grammar-based techniques. These new research directions are discussed next in detail.

7.2.1 Extending HEAD-DT's Genome: New Induction Strategies, Oblique Splits, Regression Problems

HEAD-DT can be naturally extended so its genome accounts for induction strategies other than the top-down induction. For instance, an *induction gene* could be responsible for choosing among the following approaches for tree induction: top-down induction (currently implemented), bottom-up induction (following the work of Barros et al. [1, 2] and Landeweerd et al. [10]), beam-search induction (following the work of Basgalupp et al. [3]), and possibly a hybrid approach that combines the three previous strategies.

Furthermore, one can think of a new gene to be included among the *split genes*, in which an integer would index the following split options: univariate splits (currently implemented), oblique splits (along with several parameter values that would determine the strategy for generating the oblique split), and omni splits (real-time decision for each split as whether it should be univariate or oblique, following the work of Yildiz and Alpaydin [13]).

Finally, HEAD-DT could be adapted to regression problems. For such, split measures specially designed to regression have to be implemented—see, for instance, Chap. 2, Sect. 2.3.1.

7.2.2 Multi-objective Fitness Function

In Chap. 6, we tested 15 different single-objective fitness functions for HEAD-DT. A natural extension regarding fitness evaluation would be to employ multi-objective strategies to account for multiple objectives being simultaneously optimised. For instance, instead of searching only for the largest average F-Measure achieved by a candidate algorithm in the meta-training set, HEAD-DT could look for an algorithm that induces trees with reduced size. Considering that, for similar predictive performance, a simpler model is always preferred (as stated by the well-known Occam's razor principle), it makes sense to optimise both a measure of predictive performance (such as F-Measure) and a measure of model complexity (such as tree size).

Possible solutions for dealing with multi-objective optimisation include the Pareto dominance approach and the lexicographic analysis [8]. The first assumes the set of non-dominated solutions is provided to the user (instead of a single best solution). Hence, the evolutionary algorithm must be modified as to properly handle the selection operation, as well as elitism and the return of the final solution. A lexicographic approach, in turn, assumes that each objective has a different priority order, and thus it decides which individual is best by traversing the objectives from the highest to the lowest priority. Each multi-objective strategy has advantages and disadvantages, and an interesting research effort would be to compare a number of strategies, so one could see how they cope with different optimisation goals.

7.2.3 Automatic Selection of the Meta-Training Set

We employed a methodology that randomly selected data sets to be part of the meta-training set. Since the performance of the evolved decision-tree algorithm is directly related to the data sets that belong to the meta-training set, we believe an intelligent and automatic strategy to select a proper meta-training set would be beneficial to the final user. For instance, the user could provide HEAD-DT with a list of available data sets, and HEAD-DT would automatically select from this list those data sets which are more similar to the available meta-test set.

Some possibilities for performing this intelligent automatic selection include clustering or selecting the k-nearest-neighbors based on meta-features that describe these data sets, i.e., selecting those data sets from the list that have largest similarity according to a given similarity measure (e.g., Euclidean distance, Mahalanobis distance, etc.). In such an approach, the difficulty lies in choosing a set of meta-features that characterize the data sets in such a way that data sets with similar meta-features require similar design components of a decision-tree algorithm. This is, of course, not trivial, and an open problem in the meta-learning research field [4].

7.2.4 Parameter-Free Evolutionary Search

One of the main challenges when dealing with evolutionary algorithms is the large amount of parameters one needs to set prior to its execution. We "avoided" this problem by consistently employing typical parameter values found in the literature of evolutionary algorithms for decision-tree induction. Nevertheless, we still tried to optimise the parameter p in HEAD-DT, which controls the probabilities of crossover, mutation, and reproduction during evolution.

A challenging research effort would be the design of self-adaptive evolutionary parameters, which dynamically detect when their values need to change for the sake of improving the EA's performance. Research in that direction includes the work of Kramer [9], which proposes self-adaptive crossover points, or the several studies on EAs for tuning EAs [6].

7.2.5 Solving the Meta-Overfitting Problem

The meta-overfitting problem presented in Sect. 5.2.3 offers a good opportunity for future work. Recall that meta-overfitting occurs when we want to generate a good "all around" algorithm, i.e., an algorithm that is robust across a set of different data sets. Some alternatives we envision for solving (or alleviating) this problem are:

- **Increase the number of data sets in the meta-training set**. Since the goal is to generate an "all-around" algorithm that performs well regardless of any particular data set characteristic, it is expected that feeding HEAD-DT with a larger meta-training set would increase the chances of achieving this goal. The disadvantage of this approach is that HEAD-DT becomes increasingly slower with each new data set that is added to the meta-training set.
- **Build a replacing mechanism during the evolutionary process that dynamically updates the data sets in the meta-training set**. By feeding HEAD-DT with different data sets per generation, we could guide the evolution towards robust "all-around" algorithms, avoiding the extra computational time spent in the previous solution. The problem with this strategy is that HEAD-DT would most likely provide an algorithm that performs well in the meta-training set used in the last generation of the evolution. This could be fixed by storing the best algorithm in each generation and then executing a single generation with these algorithms in the population, and with a full meta-training set (all data sets used in all generations).
- **Build a meta-training set with *diversity***. A good "all-around" algorithm should perform reasonably well in all kinds of scenarios, and thus a possible solution is to feed HEAD-DT with data sets that cover a minimum level of diversity in terms of structural characteristics, that in turn represent a particular scenario. In practice, the problem with this approach is to identify the characteristics that really influence in predictive performance. As previously discussed, this is an open problem and a research area by itself, known as *meta-learning*.

7.2.6 Ensemble of Automatically-Designed Algorithms

The strategy for automatically designing decision-tree induction algorithms presented in this book aimed at the generation of effective algorithms, which were capable of outperforming other traditional manually-designed decision-tree algorithms.

Another promising research direction would be to automatically design decision-tree algorithms to be used in an ensemble of classifiers. In this case, each individual would be an ensemble of automatically-designed algorithms, and a multi-objective fitness function could be employed to account for both the ensemble's predictive performance in a meta-training set and its diversity regarding each automatically-designed algorithm. For measuring diversity, one could think of a measure that would take into account the number of distinct correctly-classified instances between two different algorithms.

7.2.7 Grammar-Based Genetic Programming

Finally, a natural extension of HEAD-DT would be regarding its search mechanism: instead of relying on a GA-like evolutionary algorithm, more sophisticated EAs such as standard grammar-based genetic programming (GGP) or grammatical evolution (GE) could be employed to evolve candidate decision trees. The latter seems to be an interesting research direction, since it is nowadays one of the most widely applied genetic programming methods.

References

1. R.C. Barros et al., A bottom-up oblique decision tree induction algorithm, in *11th International Conference on Intelligent Systems Design and Applications*. pp. 450–456 (2011)
2. R.C. Barros et al., A framework for bottom-up induction of decision trees, in Neurocomputing (in press 2013)
3. M.P. Basgalupp et al., A beam-search based decision-tree induction algorithm, in *Machine Learning Algorithms for Problem Solving in Computational Applications: Intelligent Techniques. IGI-Global* (2011)
4. P. Brazdil et al., *Metalearning—Applications to Data Mining*. Cognitive Technologies (Springer, Berlin, 2009), pp. I-X, 1–176. ISBN: 978-3-540-73262-4
5. L. Breiman et al., *Classification and Regression Trees* (Wadsworth, Belmont, 1984)
6. A.E. Eiben, S.K. Smit, Parameter tuning for configuring and analyzing evolutionary algorithms. Swarm Evol. Comput. **1**(1), 19–31 (2011)
7. A. Frank, A. Asuncion. *UCI Machine Learning Repository* (2010)
8. A.A. Freitas, A critical review of multi-objective optimization in data mining: a position paper. SIGKDD Explor. Newsl. **6**(2), 77–86 (2004). ISSN: 1931–0145
9. O. Kramer, Self-Adaptive Crossover, *Self-Adaptive Heuristics for Evolutionary Computation.*, Vol. 147. Studies in Computational Intelligence (Springer, Berlin, 2008)
10. G. Landeweerd et al., Binary tree versus single level tree classification of white blood cells. Pattern Recognit. **16**(6), 571–577 (1983)
11. J.R. Quinlan, *C4.5: Programs for Machine Learning* (Morgan Kaufmann, San Francisco, 1993). ISBN: 1-55860-238-0
12. M. Souto et al., Clustering cancer gene expression data: a comparative study. BMC Bioinform. **9**(1), 497 (2008)
13. C.T. Yildiz, E. Alpaydin, Omnivariate decision trees. IEEE Trans. Neural Netw. **12**(6), 1539–1546 (2001)